STANDARD GRADE | GENERAL

2008

[BLANK PAGE]

FOR OFFICIAL USE

KU   PS

Total
Marks

G

0500/401

NATIONAL
QUALIFICATIONS
2008

THURSDAY, 1 MAY
9.00 AM – 10.30 AM

CHEMISTRY
STANDARD GRADE
General Level

Fill in these boxes and read what is printed below.

Full name of centre

Town

Forename(s)

Surname

Date of birth
Day Month Year   Scottish candidate number   Number of seat

1 All questions should be attempted.

2 Necessary data will be found in the Data Booklet provided for Chemistry at Standard Grade and Intermediate 2.

3 The questions may be answered in any order but all answers are to be written in this answer book, and must be written clearly and legibly in ink.

4 Rough work, if any should be necessary, as well as the fair copy, is to be written in this book.

   Rough work should be scored through when the fair copy has been written.

5 Additional space for answers and rough work will be found at the end of the book.

6 The size of the space provided for an answer should not be taken as an indication of how much to write. It is not necessary to use all the space.

7 Before leaving the examination room you must give this book to the invigilator. If you do not, you may lose all the marks for this paper.

PART 1

In Questions 1 to 9 of this part of the paper, an answer is given by circling the appropriate letter (or letters) in the answer grid provided.

In some questions, two letters are required for full marks.

If more than the correct number of answers is given, marks will be deducted.

A total of 20 marks is available in this part of the paper.

SAMPLE QUESTION

A CH_4	B H_2	C CO_2
D CO	E C_2H_5OH	F C

(a) Identify the hydrocarbon.

Ⓐ	B	C
D	E	F

The one correct answer to part (a) is A. This should be circled.

(b) Identify the **two** elements.

A	Ⓑ	C
D	E	Ⓕ

As indicated in this question, there are **two** correct answers to part (b). These are B and F. Both answers are circled.

If, after you have recorded your answer, you decide that you have made an error and wish to make a change, you should cancel the original answer and circle the answer you now consider to be correct. Thus, in part (a), if you want to change an answer A to an answer D, your answer sheet would look like this:

Ⱥ	B	C
Ⓓ	E	F

If you want to change back to an answer which has already been scored out, you should enter a tick (✓) in the box of the answer of your choice, thus:

✓Ⱥ	B	C
Ꝺ	E	F

Marks | KU | PS

1. The Periodic Table shows the names of the elements.

A nitrogen	B lithium	C aluminium
D sodium	E oxygen	F platinum

(a) Identify the **two** elements which have similar chemical properties.

You may wish to use page 8 of the data booklet to help you.

A	B	C
D	E	F

1

(b) Identify the element discovered in 1807.

You may wish to use page 8 of the data booklet to help you.

A	B	C
D	E	F

1

(c) Identify the element which is used as the catalyst in the Ostwald Process.

A	B	C
D	E	F

1

(d) Identify the **two** elements which form a covalent compound.

A	B	C
D	E	F

1

(4)

[Turn over

Marks KU PS

2. The grid shows the names of some elements.

A	hydrogen
B	helium
C	oxygen
D	silicon
E	carbon

(a) Identify the **two** elements which exist as **diatomic** molecules.

A
B
C
D
E

1

(b) Identify the element which has the electron arrangement 2,4.

You may wish to use page 1 of the data booklet to help you.

A
B
C
D
E

1

(c) Identify the element which must be present for iron to rust.

A
B
C
D
E

1

(3)

Marks | KU | PS

3. Electricity can be produced using electrochemical cells.

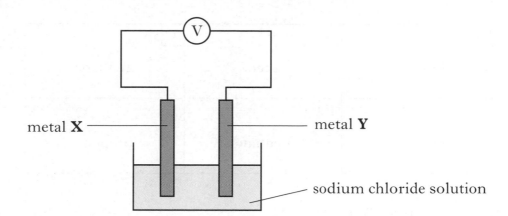

	metal X	metal Y
A	copper	lead
B	copper	magnesium
C	copper	copper
D	copper	nickel

(a) Identify the arrangement which would **not** produce electricity.

A

B

C

D

1

(b) Identify the arrangement which would produce the **largest** voltage.
You may wish to use page 7 of the data booklet to help you.

A

B

C

D

1

(2)

[Turn over

Marks | KU | PS

4. The names of some hydrocarbons are shown in the grid.

A	B	C
butene	ethene	methane
D	E	F
hexene	pentane	propene

(a) Identify the **two** alkanes.

A	B	C
D	E	F

1

(b) Identify the hydrocarbon with a boiling point of 36 °C.
You may wish to use page 6 of the data booklet to help you.

A	B	C
D	E	F

1

(c) Identify the hydrocarbon with molecular formula C_4H_8.

A	B	C
D	E	F

1

(3)

5. Coating iron prevents rusting.

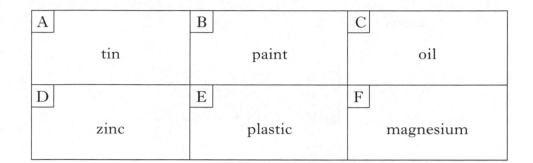

A	B	C
tin	paint	oil
D	E	F
zinc	plastic	magnesium

(a) Identify the coating which is used to galvanise iron.

A	B	C
D	E	F

1

(b) Identify the coating, which, if scratched, would cause the iron to rust **fastest**.

You may wish to use page 7 of the data booklet to help you.

A	B	C
D	E	F

1

(2)

[Turn over

Marks | KU | PS

6. A student carried out an experiment to investigate the viscosity of different oils.

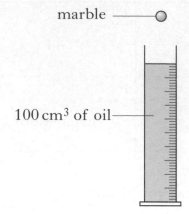

marble

100 cm^3 of oil

He timed how long it took for a marble to fall through 100 cm^3 of each oil fraction.

His results are shown in the table.

Oil	Time/s
1	6
2	10
3	15
4	23

Identify the **correct** statement.

A	Oil 1 is most viscous.
B	Oil 4 is least viscous.
C	Oil 2 is more viscous than oil 3.
D	Oil 4 is more viscous than oil 1.

(1)

Marks KU PS

7. The grid shows the names of some chlorides.

A	B	C
calcium chloride	barium chloride	magnesium chloride
D	E	F
sodium chloride	silver chloride	potassium chloride

(a) Identify the chloride which could be produced by a precipitation reaction.

You may wish to use page 5 of the data booklet to help you.

A	B	C
D	E	F

1

(b) Identify the chloride which could be used as a fertiliser.

A	B	C
D	E	F

1

(2)

[Turn over

Marks | KU | PS

8. Different terms can be used to indicate the number of atoms in a molecule.

	Term	Number of atoms in a molecule
A	tri-atomic	3
B	tetra-atomic	4
C	penta-atomic	5
D	hexa-atomic	6

Identify the term used to describe the following molecule.

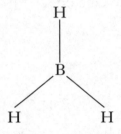

A
B
C
D

(1)

DO NOT
WRITE IN
THIS
MARGIN

Marks | KU | PS

9. A technician set up an experiment to investigate electrical conductivity.

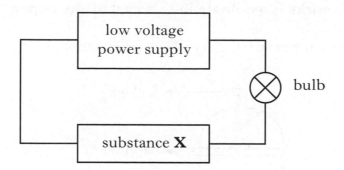

	Substance X
A	molten metal
B	covalent liquid
C	ionic solution
D	ionic solid
E	solid metal

Identify the **two** experiments in which the bulb would **not** light.

A
B
C
D
E

(2)

[Turn over

PART 2

A total of 40 marks is available in this part of the paper.

10. The diagram shows a tower in which crude oil is separated.

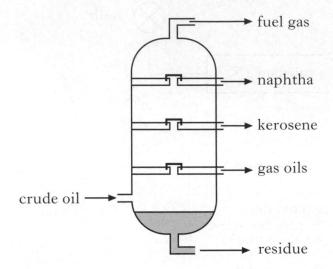

(a) Name the process used to separate crude oil.

_____   1

(b) Naphtha can be cracked to produce molecules that are more useful.

How does the **size** of these more useful molecules compare to the **size** of the molecules in naphtha?

_____

_____   1

(c) In industry the catalyst used to crack naphtha is zeolite.
Zeolite is a substance that contains aluminium silicate.

Name the elements present in aluminium silicate.

_____   1

(3)

Marks | KU | PS

11. A teacher demonstrated the following experiment.

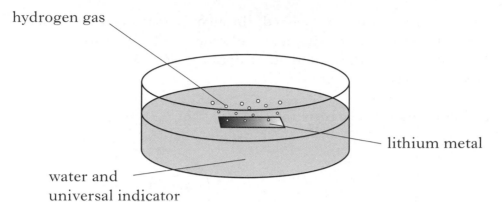

hydrogen gas

lithium metal

water and
universal indicator

(a) State the test for hydrogen gas.

_____

_____   **1**

(b) The universal indicator turned purple.

Circle the correct word to complete the sentence.

A solution which turns universal indicator purple is $\left\{ \begin{array}{l} \text{acidic} \\ \text{neutral} \\ \text{alkaline} \end{array} \right\}$.   **1**

(c) Why are metals, like lithium, stored under oil?

_____

_____   **1**

(3)

[Turn over

12.   **Manufacture of Titanium**

Carbon and titanium oxide are passed through a reactor to produce carbon monoxide and impure titanium chloride. The impurities are removed by distillation. Pure titanium chloride reacts with sodium to produce titanium and sodium chloride.

(*a*) Use the information to complete the flow diagram.

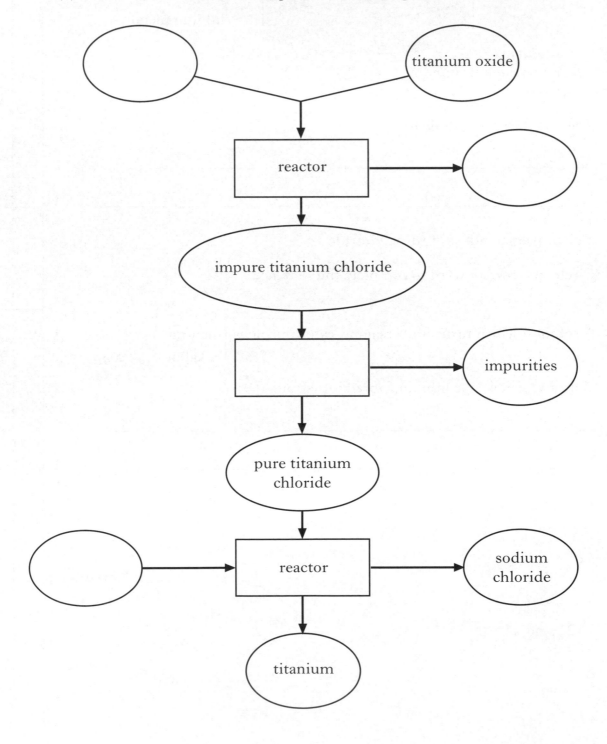

2

Marks

12. **(continued)**

(b) Titanium can be mixed with other metals to make a substance that is strong and lightweight.

What term is used to describe a mixture of metals?

_____

1

(c) Medical instruments can be made from a mixture of metals containing 76% titanium, 4% zirconium and the rest is other metals.

Label the pie chart to show the name and percentage for each part of the mixture.

(An additional pie chart, if required, can be found on page 28.)

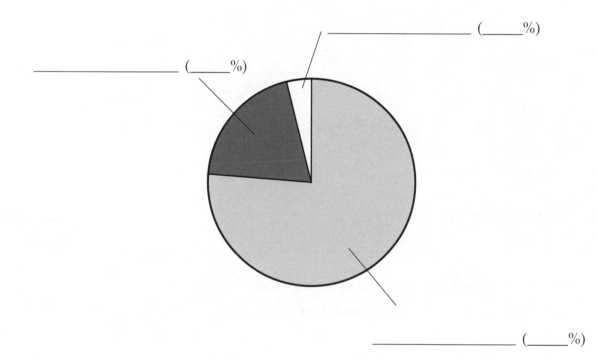

_____ (____%)

_____ (____%)

_____ (____%)

2

(5)

[**Turn over**

Marks | KU | PS

13. A teacher demonstrated the following experiment.

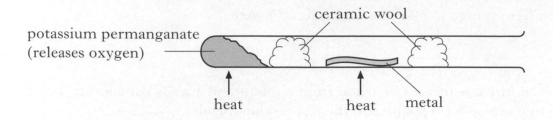

potassium permanganate
(releases oxygen)

ceramic wool

heat   heat   metal

Her results are shown in the table below.

Metal	Observation
zinc	glowed brightly
copper	dull red glow
silver	no reaction

(*a*) (i) Predict what would be seen if the experiment was repeated using magnesium.

You may wish to use page 7 of the data booklet to help you.

_____

_____   1

(ii) The experiment was repeated using **powdered** zinc.

How would this affect the **speed** of the reaction?

_____

_____   1

(*b*) Silver is found uncombined in the Earth's crust.

Name another metal which is found uncombined in the Earth's crust.

You may wish to use page 7 of the data booklet to help you.

_____   1

(3)

DO NOT WRITE IN THIS MARGIN

Marks KU PS

14. Flowers produce a sweet-tasting liquid called nectar.

Nectar contains a mixture of sugars such as glucose and sucrose.

(*a*) To which family of compounds do glucose and sucrose belong?

____ **1**

(*b*) Glucose can be broken down to produce alcohol.

(i) Name this **type** of chemical reaction.

____ **1**

(ii) What is the chemical name for the alcohol produced?

____ **1**

(3)

[Turn over

Marks   KU   PS

15. The table below shows the mass of some ions found in a 1 litre sample of water.

Ion	Mass/mg
chloride	10
sulphate	50
calcium	70
magnesium	15
potassium	4

(*a*)   Present the information as a bar chart.

Use appropriate scales to fill most of the graph paper.

(Additional graph paper, if required, can be found on page 28.)

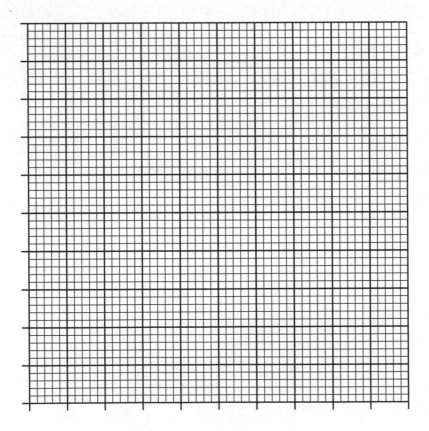

2

Marks | KU | PS

15. **(continued)**

(*b*) The bicarbonate ion is also present in the sample of water.

When heated the bicarbonate ion breaks down to form carbon dioxide gas.

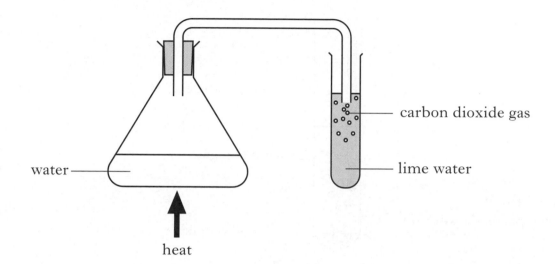

water

heat

carbon dioxide gas

lime water

    (i) Write the formula for carbon dioxide gas.

         _____

1

    (ii) Describe what would be seen when carbon dioxide gas is bubbled through lime water.

         _____

         _____

1

(4)

[Turn over

Marks | KU | PS

16. A student investigated the amount of the biological catalyst, catalase, in different vegetables.

Catalase breaks down hydrogen peroxide solution to produce water and oxygen.

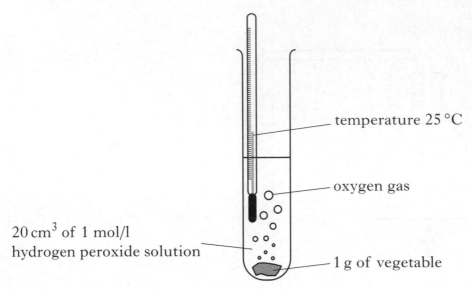

temperature 25 °C

oxygen gas

20 cm³ of 1 mol/l hydrogen peroxide solution

1 g of vegetable

The results are shown in the table.

Vegetable	Number of bubbles of oxygen gas in 3 minutes
leek	40
potato	10
parsnip	65
horseradish	5

(a) Using the information in the table, name the vegetable which contains the largest amount of catalase.

_____

1

(b) What term is used to describe a biological catalyst such as catalase?

_____

1

16. (continued)

(*c*) The experiment was repeated to find out if increasing the concentration of hydrogen peroxide solution would speed up the reaction.

Complete the labelling of the diagram to show how she would make her second experiment a fair test.

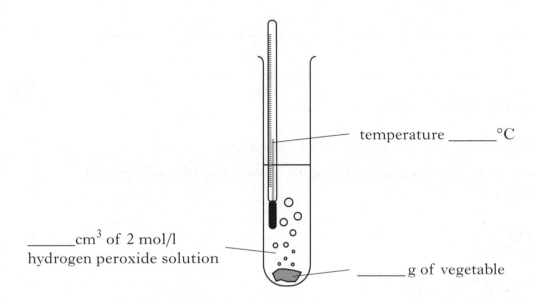

temperature _____°C

_____cm^3 of 2 mol/l
hydrogen peroxide solution

_____ g of vegetable

1

(3)

[Turn over

Marks   KU   PS

17. The plastic poly(chloroethene) has many uses.

(a) Name the monomer used to make poly(chloroethene).

_____

1

(b) Poly(chloroethene) is **non**-biodegradable.
State why this may be an **advantage**.

_____

_____

_____

1

(c) Poly(chloroethene) can be used as a fibre in clothing.

A student used the apparatus shown to investigate the strength of different fibres.

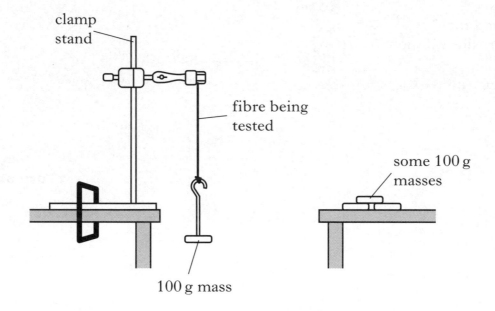

17. **(c)** **(continued)**

His results are shown in the table.

Fibre	Mass to break fibres/g
cotton	600
polyester	1200
wool	200
poly(chloroethene)	1000
poly(propene)	1100

(i) How does the strength of the synthetic fibres compare to the strength of the natural fibres?

_____

_____   1

(ii) He tested another fibre and found that the mass needed to break it was 300 g.

Predict whether this fibre is natural or synthetic.

_____   1

(4)

[Turn over

Marks   KU   PS

18.   Crude oil contains sulphur compounds, such as hydrogen sulphide.

(*a*)   Hydrogen sulphide burns in oxygen to produce sulphur dioxide and water.

Write a **word** equation for this reaction.

1

(*b*)   The sulphur dioxide produced is used to manufacture sulphuric acid. Part of the manufacture of sulphuric acid is shown.

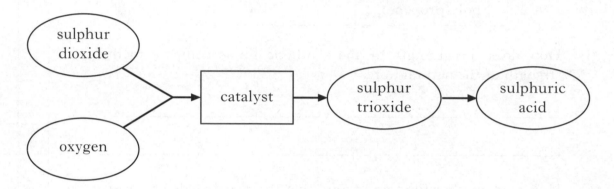

(i)   What is the purpose of a catalyst?

_____

_____   1

(ii)   The table shows the percentage of sulphur trioxide produced at different temperatures.

Temperature of catalyst/°C	Percentage of sulphur trioxide produced
442	99·5
475	95·0
518	88·0
600	63·0

What effect does increasing the temperature of the catalyst have on the percentage of sulphur trioxide produced?

_____

_____   1

(3)

19. Rechargeable batteries are used in cars.

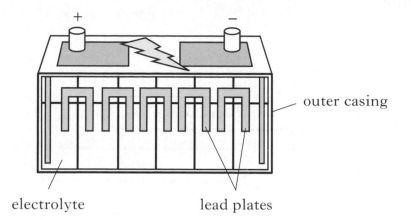

(a) Name the electrolyte used in a car battery.

_____   1

(b) A car battery has six cells joined together.

The voltage of the car battery is **12 volts**.

What is the voltage of **one** cell in the car battery?

_____ volts   1

(c) Some cars use the fuel "LPG" rather than petrol.

What is meant by the term **fuel**?

_____

_____   1

(d) "LPG" is a mixture of hydrocarbons.

Name the **two** compounds produced when "LPG" burns in a plentiful supply of air.

_____   1

(4)

[Turn over

Marks | KU | PS

20. The chart shows the pH range of soil in which different vegetables can grow successfully.

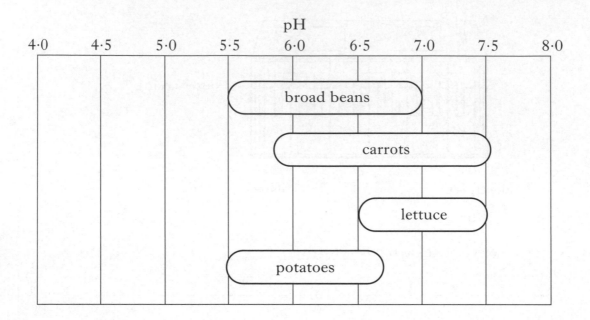

(a) The soil in a garden has a pH of 6·0.

Name the vegetable which would **not** grow successfully in this garden.

_____

1

(b) Another garden has soil pH of 4·5.

Name a substance that could be added to the soil in order to grow all the vegetables successfully.

_____

1

(2)

DO NOT WRITE IN THIS MARGIN

Marks KU PS

21. Acids have many uses.

(*a*) Phosphoric acid is found in a fizzy drink.

Suggest the pH of the fizzy drink.

____________________ **1**

(*b*) Nitric acid can be used to make fertilisers.

Explain why there has been a major increase in the use of fertilisers over the last 100 years.

____

____

____ **1**

(*c*) Dilute hydrochloric acid reacts with zinc metal.

The equation for the reaction is:

hydrochloric acid + zinc $\longrightarrow$ compound **X** + hydrogen

Name compound **X**.

________________________________ **1**

(3)

[*END OF QUESTION PAPER*]

ADDITIONAL SPACE FOR ANSWERS

ADDITIONAL PIE CHART FOR QUESTION 12(*c*)

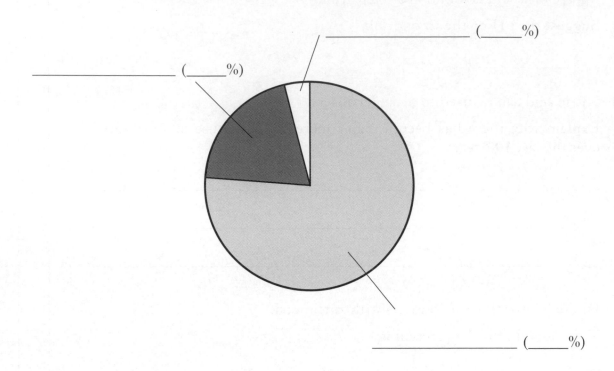

_____ (____%)

_____ (____%)

_____ (____%)

ADDITIONAL GRAPH PAPER FOR QUESTION 15(*a*)

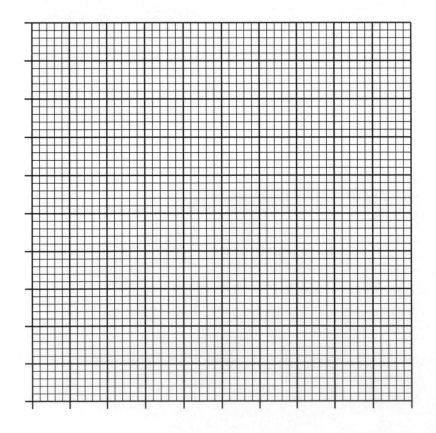

ADDITIONAL SPACE FOR ANSWERS

ADDITIONAL SPACE FOR ANSWERS

ADDITIONAL SPACE FOR ANSWERS

ADDITIONAL SPACE FOR ANSWERS

[BLANK PAGE]

FOR OFFICIAL USE

G

KU	PS

Total Marks

0500/401

NATIONAL QUALIFICATIONS 2009

MONDAY, 11 MAY 9.00 AM – 10.30 AM

CHEMISTRY STANDARD GRADE General Level

Fill in these boxes and read what is printed below.

Full name of centre

Town

Forename(s)

Surname

Date of birth
Day Month Year

Scottish candidate number

Number of seat

1 All questions should be attempted.

2 Necessary data will be found in the Data Booklet provided for Chemistry at Standard Grade and Intermediate 2.

3 The questions may be answered in any order but all answers are to be written in this answer book, and must be written clearly and legibly in ink.

4 Rough work, if any should be necessary, as well as the fair copy, is to be written in this book.

Rough work should be scored through when the fair copy has been written.

5 Additional space for answers and rough work will be found at the end of the book.

6 The size of the space provided for an answer should not be taken as an indication of how much to write. It is not necessary to use all the space.

7 Before leaving the examination room you must give this book to the invigilator. If you do not, you may lose all the marks for this paper.

PART 1

In Questions 1 to 8 of this part of the paper, an answer is given by circling the appropriate letter (or letters) in the answer grid provided.

In some questions, two letters are required for full marks.

If more than the correct number of answers is given, marks will be deducted.

A total of 20 marks is available in this part of the paper.

SAMPLE QUESTION

A CH_4	B H_2	C CO_2
D CO	E C_2H_5OH	F C

(a) Identify the hydrocarbon.

Ⓐ	B	C
D	E	F

The one correct answer to part (a) is A. This should be circled.

(b) Identify the **two** elements.

A	Ⓑ	C
D	E	Ⓕ

As indicated in this question, there are **two** correct answers to part (b). These are B and F. Both answers are circled.

If, after you have recorded your answer, you decide that you have made an error and wish to make a change, you should cancel the original answer and circle the answer you now consider to be correct. Thus, in part (a), if you want to change an answer A to an answer D, your answer sheet would look like this:

Ⓐ̶	B	C
Ⓓ	E	F

If you want to change back to an answer which has already been scored out, you should enter a tick (✓) in the box of the answer of your choice, thus:

✓Ⓐ̶	B	C
D̶	E	F

1. The diagram shows part of the Periodic Table.

The letters do **not** represent the symbols for the elements.

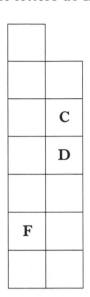

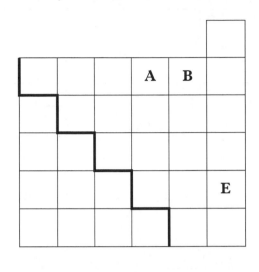

(a) Identify the element which has the electron arrangement 2, 7.

You may wish to use page 1 of the data booklet to help you.

A	B	C	D	E	F

1

(b) Identify the unreactive element.

A	B	C	D	E	F

1

(c) Identify the **two** elements which are in the same group.

A	B	C	D	E	F

1

(3)

[Turn over

Marks | KU | PS

2. The grid contains the names of some metals.

A mercury	B magnesium	C copper
D iron	E silver	F sodium

(a) Identify the metal used as the catalyst in the Haber Process.

A	B	C
D	E	F

1

(b) Identify the metal with the highest density.
You may wish to use page 2 of the data booklet to help you.

A	B	C
D	E	F

1

(c) Identify the metal which was discovered after 1790.
You may wish to use page 8 of the data booklet to help you.

A	B	C
D	E	F

1

(3)

Marks | KU | PS

3. Two students investigated the reaction between magnesium and dilute hydrochloric acid.

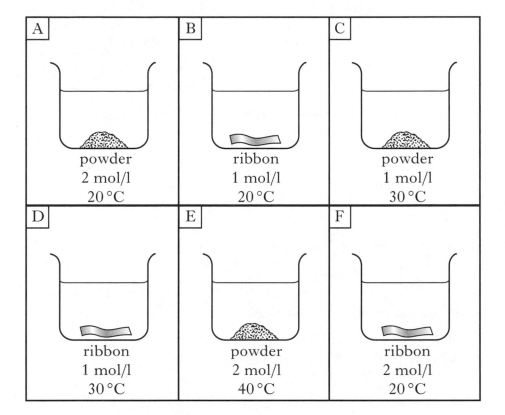

(a) Identify the **two** experiments which could be used to show the effect of concentration on the speed of reaction.

A	B	C
D	E	F

1

(b) Identify the experiment with the fastest speed of reaction.

A	B	C
D	E	F

1

(2)

[Turn over

4. The grid contains the names of some gases.

A argon	B hydrogen	C oxygen
D nitrogen	E carbon dioxide	F ammonia

(a) Identify the gas which makes up approximately 20% of air.

A	B	C
D	E	F

1

(b) Identify the gas which burns with a pop.

A	B	C
D	E	F

1

(c) Identify the gas produced during respiration.

A	B	C
D	E	F

1

(d) Identify the gas produced when a metal reacts with dilute acid.

A	B	C
D	E	F

1

(4)

DO NOT WRITE IN THIS MARGIN

Marks KU PS

5. The grid contains the names of some elements.

A hydrogen	B carbon	C copper
D zinc	E nitrogen	F iron

(*a*) Identify the element produced in a blast furnace. **1**

A	B	C
D	E	F

(*b*) Identify the element used to galvanise steel objects. **1**

A	B	C
D	E	F

(2)

[Turn over

DO NOT WRITE IN THIS MARGIN

Marks KU PS

6. The grid shows the names of some compounds.

A lead sulphate	B sodium chloride
C calcium hydroxide	D potassium phosphate

(*a*) Identify the compound which contains only **two** elements.

A	B
C	D

1

(*b*) Identify the compound which will neutralise an acid.

A	B
C	D

1

(2)

DO NOT WRITE IN THIS MARGIN

Marks KU PS

7. A student made some statements about glucose.

A	Glucose is a carbohydrate.
B	Glucose is insoluble in water.
C	Glucose is made during photosynthesis.
D	Iodine solution can be used to test for glucose.
E	Glucose molecules are too large to pass through the gut wall.

Identify the **two** correct statements.

A
B
C
D
E

(2)

[Turn over

Marks | KU | PS

8. Silver and gold are used to make jewellery.

Identify the **two** statements which are true for **both** silver and gold.

You may wish to use the data booklet to help you.

A	They are transition metals.
B	They do **not** conduct electricity.
C	They are more reactive than lead.
D	They react with hydrochloric acid.
E	They are found uncombined in the Earth's crust.

A
B
C
D
E

(2)

PART 2

A total of 40 marks is available in this part of the paper.

9. Coal is a fossil fuel which burns, releasing heat energy.

(a) What name is given to all chemical reactions which release heat energy?

_____

1

(b) Coal is a finite resource.

What is meant by the term **finite**?

_____

_____

1

(c) Name another fossil fuel.

_____

1

(3)

[Turn over

Marks | KU | PS

10. Magnesium and chlorine are common elements.

(*a*) Complete the table.

You may wish to use page 8 of the data booklet to help you.

Element	Atomic Number	Metal or non-metal
magnesium		
chlorine		

1

(*b*) Magnesium and chlorine react together to form magnesium chloride.

(i) Write the formula for magnesium chloride.

1

(ii) Using information from page 6 of the data booklet, enter the melting point and boiling point of magnesium chloride on the diagram below.

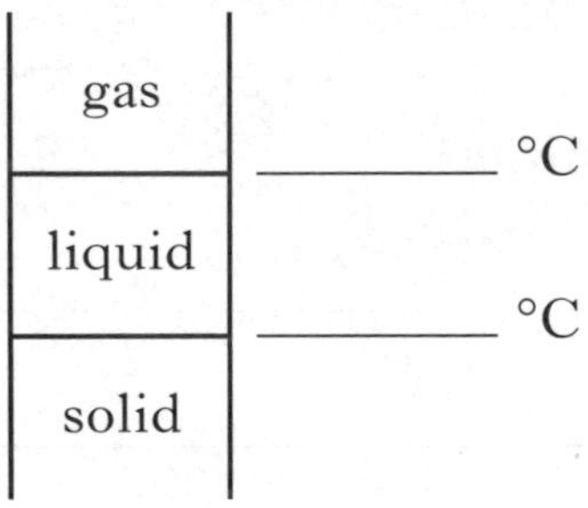

1

(iii) ⟨Circle⟩ the correct word to complete the following sentence.

At 1000 °C magnesium chloride is a
$\left\{ \begin{array}{l} \text{gas} \\ \text{liquid} \\ \text{solid} \end{array} \right\}$.

1

(4)

11. Plastics have many uses. Perspex is used to make advertising signs. Artificial limbs can be made from PVC. Polythene can be used to make carrier bags and egg cartons can be made from polystyrene.

(a) Present this information as a table with suitable headings.

2

(b) Scientists have produced a plastic which is biodegradable.

What is meant by the term **biodegradable**?

_____

_____   1

(c) PVC softens when heated and can be easily reshaped.

What term is used to describe this type of plastic?

_____   1

(d) Name the monomer which is used to make polystyrene.

_____   1

(e) Name the type of chemical reaction which is used to make polystyrene.

_____   1

(6)

[Turn over

Marks | KU | PS

12. A student carried out the experiment shown.

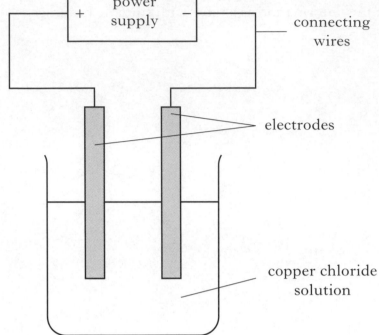

(a) Name the product formed at the positive electrode.

_____   1

(b) Name the charged particles that flow through the connecting wires.

_____   1

(c) Name a non-metal element which is suitable for use as the electrodes.

_____   1

(3)

Marks KU PS

13. Long chain alkanes in diesel can be broken down using aluminium oxide as a catalyst.

ceramic wool
soaked in
diesel

0·8 g
aluminium oxide

heat

bromine
water
decolourises

(*a*) What mass of aluminium oxide will be present at the end of the experiment?

_____ g

1

(*b*) One reaction taking place in the heated test tube is:

$$C_{16}H_{34} \longrightarrow C_7H_{14} + C_9H_{20}$$

(i) Name this type of chemical reaction.

_____

1

(ii) On the equation above circle the formula of the product which decolourised the bromine water.

1

(3)

[Turn over

14.   The pie chart shows the uses of oxygen.

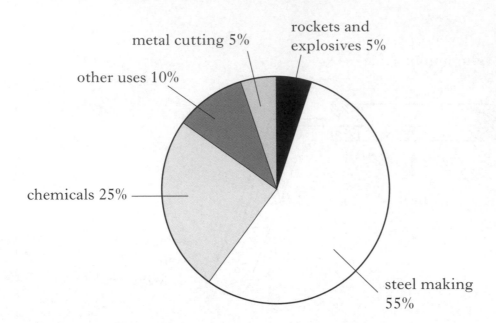

metal cutting 5%

rockets and explosives 5%

other uses 10%

chemicals 25%

steel making 55%

(*a*)   Present the information as a bar chart.

Use appropriate scales to fill most of the graph paper.

(Additional graph paper, if required, can be found on page 23.)

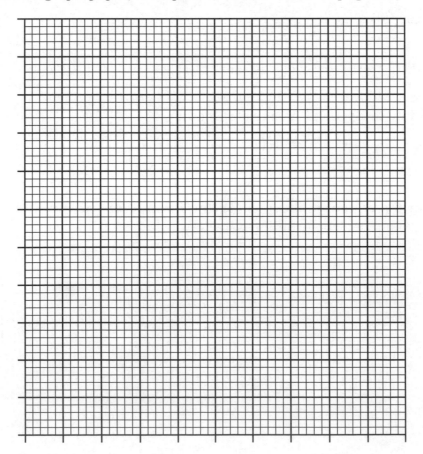

2

14. (continued)

(b) Oxygen is made up of diatomic molecules.

What is meant by the term **diatomic**?

_____

_____

_____   **1**

(c) Steel is a mixture of metals.

What name is given to a mixture of metals such as steel?

_____   **1**

(d) Different methods can be used to prevent steel from rusting.

   (i) How does tin-plating prevent rusting?

     _____

     _____   **1**

   (ii) Name a metal which can be used to provide sacrificial protection to steel.

     _____   **1**

(6)

[Turn over

Marks | KU | PS

15. Batteries are used in a range of items. A battery is a number of cells joined together.

(a) Give a **disadvantage** in using a battery rather than mains electricity.

_____

_____   1

(b) A student investigated how different metals affect the voltage produced by a simple cell.

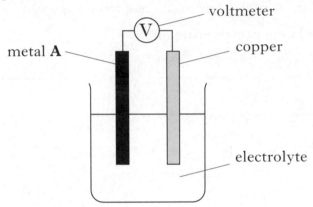

The results are shown in the table.

Metal A	Voltage/V
magnesium	2·7
tin	0·5

(i) The student set up another cell using iron and copper.

Suggest the voltage produced by this cell.

You may wish to use page 7 of the data booklet to help you.

_____ V   1

(ii) Suggest **one** factor which the student would have kept the same to make a fair comparison.

_____   1

(3)

Marks | KU | PS

16. (*a*) Ammonium nitrate is a synthetic fertiliser. It contains nitrogen which is essential for plant growth.

   (i) What is meant by the term **synthetic**?

   _____

   _____   1

   (ii) Name another essential element supplied by fertilisers.

   _____   1

(*b*) When ammonium nitrate is heated with calcium hydroxide, a colourless gas is produced. The gas turns damp pH paper blue.

ammonium nitrate
+
calcium hydroxide

damp pH paper
turns blue

heat

Name the gas produced.

_____   1

(3)

[Turn over

Marks | KU | PS

17. Yoghurt is made by fermenting fresh milk. Enzymes help to convert lactose in the milk to lactic acid.

(*a*) What is an enzyme?

_____

_____   1

(*b*) The diagram shows a molecule of lactic acid.

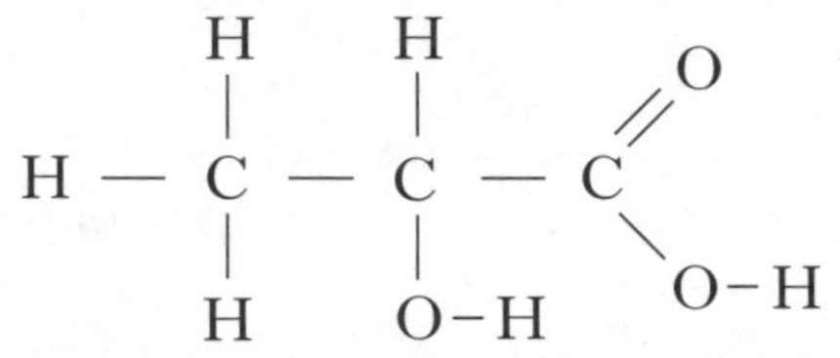

$$H - \overset{\overset{\displaystyle H}{|}}{\underset{\underset{\displaystyle H}{|}}{C}} - \overset{\overset{\displaystyle H}{|}}{\underset{\underset{\displaystyle O-H}{|}}{C}} - C\overset{\displaystyle O}{\underset{\displaystyle O-H}{}}$$

Write the molecular formula for lactic acid.

1

(*c*) Sugar can be added to sweeten yoghurt.

Suggest why sugar is added **after** the fermentation stage and not before.

_____

_____

_____   1

(3)

Marks | KU | PS

18. A student carried out an experiment to find the pH of various solutions.

Workcard

Instructions

1. Burn the element in a gas jar of oxygen.
2. Add water to the oxide formed.
3. Add 5 drops of universal indicator and shake the gas jar.
4.
5. Record the pH.

Results:

Name of oxide	pH of solution
sulphur dioxide	2
sodium oxide	13
phosphorus oxide	
aluminium oxide	could not be measured

(*a*) Instruction 4 is missing from the workcard.

What should instruction 4 tell the student to do?

_____

_____ 1

(*b*) Complete the table showing the result the student would have obtained for phosphorus oxide. 1

(*c*) Suggest why the pH of aluminium oxide could not be measured.

You may wish to use page 5 of the data booklet to help you.

_____ 1

(3)

[Turn over

Marks | KU | PS

19. The higher the octane number of a fuel the better it burns.

Number of carbon atoms	Octane number	
	alkane	alkene
4	94	98
5	62	93
6	25	85
7	0	75

(a) How does the number of carbon atoms affect the octane number of the alkanes?

_____

_____

_____   1

(b) Predict the octane number of the **alkene** with 3 carbon atoms.

_____   1

(c) In general, how does the octane number of an **alkane** compare with the octane number of the **alkene** with the same number of carbon atoms?

_____

_____

_____   1

(3)

[END OF QUESTION PAPER]

ADDITIONAL SPACE FOR ANSWERS

ADDITIONAL GRAPH PAPER FOR QUESTION 14(*a*)

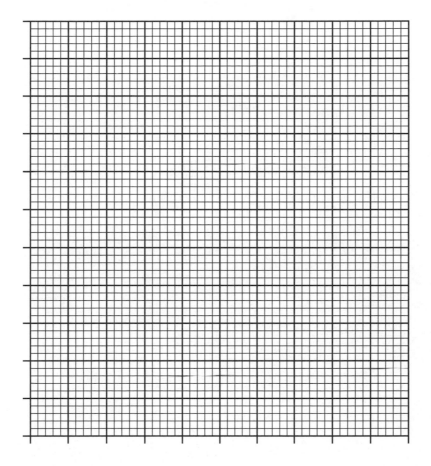

ADDITIONAL SPACE FOR ANSWERS

[BLANK PAGE]

FOR OFFICIAL USE

G

	KU	PS
Total Marks		

0500/401

NATIONAL
QUALIFICATIONS
2010

FRIDAY, 30 APRIL
9.00 AM – 10.30 AM

CHEMISTRY
STANDARD GRADE
General Level

Fill in these boxes and read what is printed below.

Full name of centre

Town

Forename

Surname

Date of birth

Day   Month   Year   Scottish candidate number   Number of seat

1　All questions should be attempted.

2　Necessary data will be found in the Data Booklet provided for Chemistry at Standard Grade and Intermediate 2.

3　The questions may be answered in any order but all answers are to be written in this answer book, and must be written clearly and legibly in ink.

4　Rough work, if any should be necessary, as well as the fair copy, is to be written in this book.

　　Rough work should be scored through when the fair copy has been written.

5　Additional space for answers and rough work will be found at the end of the book.

6　The size of the space provided for an answer should not be taken as an indication of how much to write.　It is not necessary to use all the space.

7　Before leaving the examination room you must give this book to the Invigilator.　If you do not, you may lose all the marks for this paper.

PART 1

In Questions 1 to 9 of this part of the paper, an answer is given by circling the appropriate letter (or letters) in the answer grid provided.

In some questions, two letters are required for full marks.

If more than the correct number of answers is given, marks will be deducted.

A total of 20 marks is available in this part of the paper.

SAMPLE QUESTION

A		B		C	
	CH_4		H_2		CO_2
D		E		F	
	CO		C_2H_5OH		C

(a) Identify the hydrocarbon.

Ⓐ	B	C
D	E	F

The one correct answer to part (a) is A. This should be circled.

(b) Identify the **two** elements.

A	Ⓑ	C
D	E	Ⓕ

As indicated in this question, there are **two** correct answers to part (b). These are B and F. Both answers are circled.

If, after you have recorded your answer, you decide that you have made an error and wish to make a change, you should cancel the original answer and circle the answer you now consider to be correct. Thus, in part (a), if you want to change an answer A to an answer D, your answer sheet would look like this:

A̶	B	C
Ⓓ	E	F

If you want to change back to an answer which has already been scored out, you should enter a tick (✓) in the box of the answer of your choice, thus:

✓A̶	B	C
D̶	E	F

1. The grid contains the symbols for some elements.

A	B	C
Pt	Na	P
D	E	F
N	S	Ne

(a) Identify the symbol for sodium.

You may wish to use page 8 of the data booklet to help you.

A	B	C
D	E	F

(b) Identify the element which was the first to be discovered.

You may wish to use page 8 of the data booklet to help you.

A	B	C
D	E	F

(c) Identify the symbol for a noble gas.

You may wish to use page 8 of the data booklet to help you.

A	B	C
D	E	F

1

1

1
(3)

[Turn over

DO NOT
WRITE IN
THIS
MARGIN

Marks

2. The grid contains the names of some reagents used in chemical tests.

A bromine solution	B ferroxyl indicator	C universal indicator
D lime water	E Benedict's solution	F iodine solution

(a) Identify the reagent used to test the pH of a dilute acid.

A	B	C
D	E	F

1

(b) Identify the reagent used to test for glucose.

A	B	C
D	E	F

1

(2)

3. A teacher set up some experiments to investigate the dyeing of cloth.

A — cotton — 20 °C — dye solution pH 4

B — wool — 30 °C — dye solution pH 11

C — wool — 40 °C — dye solution pH 4

D — wool — 50 °C — dye solution pH 7

E — cotton — 20 °C — dye solution pH 11

F — cotton — 30 °C — dye solution pH 7

(a) Identify the **two** experiments carried out under neutral conditions.

A	B
C	D
E	F

1

(b) Identify the **two** experiments which should be compared to show the effect of pH on the dyeing of cloth.

A	B
C	D
E	F

1

(2)

[Turn over

DO NOT WRITE IN THIS MARGIN

Marks | KU | PS

4. The grid shows the names of some oxides.

A potassium oxide	B nitrogen dioxide	C carbon monoxide
D carbon dioxide	E hydrogen oxide	F sulphur dioxide

(*a*) Identify the oxide produced by the sparking of air.

A	B	C
D	E	F

1

(*b*) Identify the oxide which dissolves in water to produce an alkaline solution.

A	B	C
D	E	F

1

(2)

DO NOT WRITE IN THIS MARGIN

Marks KU PS

5. The grid shows the formulae for some compounds.

A	B	C
$CuCl_2$	Na_2O	LiF
D	E	F
SO_2	BaF_2	$SiCl_4$

(*a*) Identify the **two** compounds which exist as molecules.

A	B	C
D	E	F

1

(*b*) Identify the compound which gives a red flame colour.
You may wish to use page 4 of the data booklet to help you.

A	B	C
D	E	F

1

(2)

[Turn over

6. The grid shows the names of different types of chemical reaction.

A neutralisation	B photosynthesis	C addition
D polymerisation	E corrosion	F combustion

(*a*) Identify the reaction in which chlorophyll absorbs light energy.

A	B	C
D	E	F

1

(*b*) Identify the reaction which takes place when iron rusts.

A	B	C
D	E	F

1

(*c*) Identify the reaction represented by the equation:

$$
\begin{array}{c}
H \quad\; H \\
|\quad\; | \\
C = C - C - H \quad + \; Br_2 \quad \longrightarrow \\
|\quad |\quad | \\
H \quad H \;\; H
\end{array}
\qquad
\begin{array}{c}
H \; Br \; H \\
|\;\;\; |\;\;\; | \\
H - C - C - C - H \\
|\;\;\; |\;\;\; | \\
Br \; H \;\; H
\end{array}
$$

A	B	C
D	E	F

1

(3)

DO NOT WRITE IN THIS MARGIN

Marks KU PS

7. The table gives information about some substances.

Substance	**Conducts as**	
	a solid	**a liquid**
A	yes	yes
B	no	yes
C	no	no
D	no	yes
E	no	no

(*a*) Identify the metal.

A
B
C
D
E

1

(*b*) Identify the **two** covalent substances.

A
B
C
D
E

1

(2)

[Turn over

DO NOT WRITE IN THIS MARGIN

Marks KU PS

8. The grid contains some statements about the effect of adding potassium hydroxide solution to dilute hydrochloric acid.

Identify the **two** correct statements.

A	Water is produced.
B	The pH of the acid decreases.
C	Hydrogen gas is produced.
D	Carbon dioxide gas is produced.
E	Potassium chloride is produced.

A
B
C
D
E

(2)

9. Identify the **two** correct statements which refer to an atom of sodium.

You may wish to use page 1 of the data booklet to help you.

A	It will form an ion by losing one electron.
B	It has two more electrons than an atom of neon.
C	It has the same atomic number as an atom of lithium.
D	It has a different size compared to an atom of bromine.
E	It has different chemical properties to an atom of potassium.

| A |
| B |
| C |
| D |
| E |

(2)

[Turn over

[BLANK PAGE]

DO NOT
WRITE IN
THIS
MARGIN

Marks | KU | PS

PART 2

A total of 40 marks is available in this part of the paper.

10. Oil and natural gas are fossil fuels.

(a) Circle the correct words to complete the sentence.

Oil was formed over $\left\{\begin{array}{c}\text{thousands}\\\text{millions}\end{array}\right\}$ of years from the remains of dead

animals and plants which decayed under the $\left\{\begin{array}{c}\text{sea bed}\\\text{land}\end{array}\right\}$.   **1**

(b) When burned, some fossil fuels produce a poisonous gas.
This gas reacts with water in the atmosphere to produce acid rain.

Name the poisonous gas.

_____   **1**

(c) Natural gas is made up mainly of methane molecules.

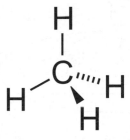

What holds the atoms together in a methane molecule?

_____

_____   **1**

(3)

[Turn over

Marks KU PS

11. Plants make glucose and oxygen gas during photosynthesis.

(a) (i) State the test for oxygen gas.

_____ 1

(ii) On the diagram below, write the names for substances **X** and **Y**.

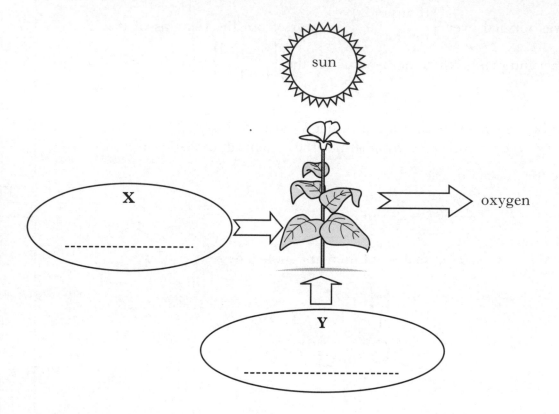

1

(b) A student set up an experiment to investigate the rate of photosynthesis in different plants.

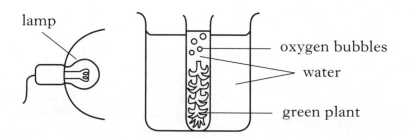

The rate of photosynthesis was measured by counting the number of bubbles of oxygen gas produced in 3 minutes.

11. (b) (continued)

The results of the investigation are shown in the table.

Name of plant	Number of bubbles of oxygen gas produced in 3 minutes
Elodea	19
Cabomba	32
Hornwort	12
Parrots Feather	24
Duckweed	8

(i) Draw a bar graph to show the information in the table.

Use appropriate scales to fill most of the graph paper.

(Additional graph paper, if required, can be found on page 25.)

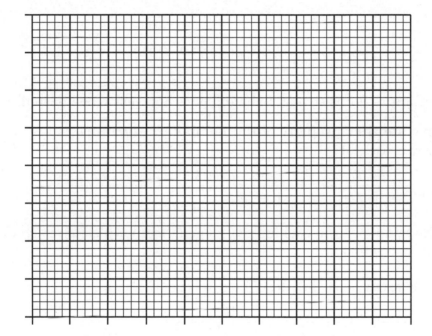

(ii) Suggest **one** factor that needs to be kept the same to make this investigation fair.

_____

2

1

(5)

[Turn over

Marks KU PS

12. Butane is a hydrocarbon which can be used as a fuel.

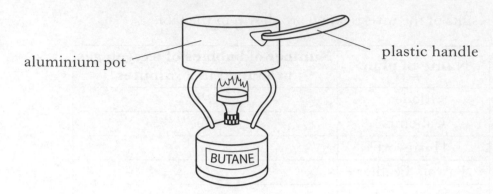

aluminium pot   plastic handle

BUTANE

(a) When butane burns in oxygen, carbon dioxide and water are produced.

Write an equation, using symbols and formulae, for the burning of butane, C_4H_{10}.

There is no need to balance the equation.

1

(b) The aluminium pot is strong, light and does not melt when heated.

(i) Give **another** property of the aluminium which makes it suitable for this use.

_____

1

(ii) The handle of the cooking pot is made from a plastic which does **not** melt when heated.

What **term** is used to describe this type of plastic?

_____

1

(c) After some time the bottom of the pot becomes covered in a black substance.

Suggest what the black substance could be.

_____

1

(4)

Marks

KU | PS

13. Potassium carbonate is a compound made up of different elements.

(a) Name the elements present in potassium carbonate.

_____ 1

(b) When potassium carbonate solution and copper nitrate solution are mixed, a chemical reaction takes place.

potassium carbonate
solution

+

copper nitrate
solution

→

solution

solid

(i) Name this type of chemical reaction.

_____ 1

(ii) Name the **solution** formed in this reaction.

You may wish to use page 5 of the data booklet to help you.

_____ 1

(iii) How could the solid be separated from the solution?

_____

_____

_____

_____

_____ 1

(4)

[Turn over

Marks

14. Iron can be mixed with other elements to produce steel for different uses.

Chromium is added to make steel suitable for use in cooking pots. Railway tracks are made from steel which contains manganese. Titanium is added to make steel suitable for aircraft parts while adding tungsten produces steel used to make hammers.

(*a*) Present the above information in a table with suitable headings.

2

Marks KU | PS

14.   (continued)

(*b*)   Steel can also be used to make storage tanks for diesel.

One method of protecting the tanks from rusting is to connect magnesium metal to them.

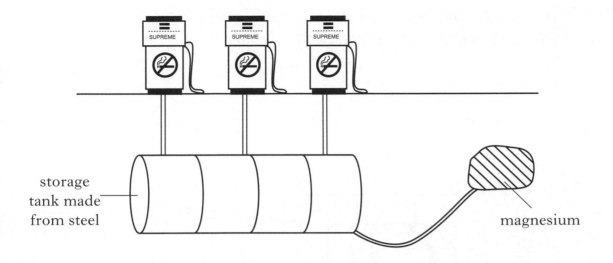

storage
tank made
from steel

magnesium

(i)   Name **two** substances which must be present for steel to rust.

_____   1

(ii)   Name the **type** of protection provided by the magnesium.

_____   1

(*c*)   Suggest another method of preventing steel from rusting.

_____   1
                                                    (5)

[Turn over

Marks   KU   P:

15. A hydrogen molecule is made up of two hydrogen atoms joined together.

(*a*) What **term** is used to describe a molecule made up of **two** atoms?

_____

1

(*b*) Hydrogen can be obtained by passing electricity through dilute acid.

What name is used to describe this process?

_____

1

(*c*) Hydrogen can be used in fuel cells to supply electricity to run a car.

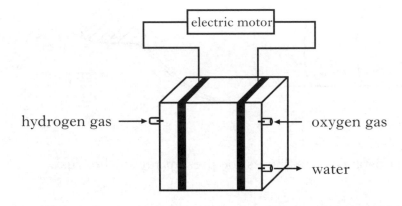

(i) Suggest **one** advantage of using fuel cells rather than petrol to power cars.

_____

_____

_____

1

(ii) Suggest a possible source of oxygen for use in the fuel cell.

_____

1

(iii) Platinum is used as the catalyst in the fuel cell.

What is the purpose of a catalyst?

_____

_____

1

(5)

16. Fertilisers are added to soil to provide essential elements required for healthy plant growth.

(a) Nitrogen is an essential element.

Name **one** other essential element required for healthy plant growth.

_____

1

(b) Some compounds containing essential elements are unsuitable for use as fertilisers.

Suggest a reason for this.

_____

_____

1

(c) Certain plants contain bacteria which can convert nitrogen from the air into nitrogen compounds.

Which part of the plant contains these bacteria?

_____

1

(3)

[Turn over

Marks   KU   PS

17. Octane is a hydrocarbon.

(*a*) To which family of hydrocarbons does octane belong?

_____   1

(*b*) The diagram shows the apparatus used to crack octane.

Octane is cracked using an aluminium oxide catalyst. Bromine solution is used to show that some of the products are unsaturated.

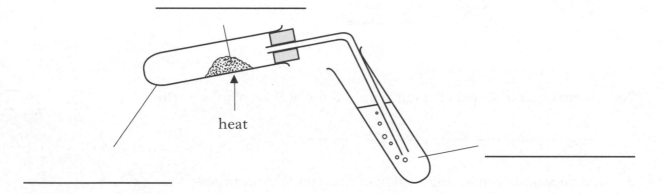

heat

(i) Label the diagram of the apparatus used to crack octane.

(An additional diagram, if required, may be found on page 25.)   1

(ii) One of the reactions taking place is:

$$C_8H_{18} \longrightarrow C_3H_6 + \mathbf{X}$$
octane

Identify **X**.

1

(iii) The product, C_3H_6 can be used to make a polymer.

Name the polymer formed when many C_3H_6 molecules join together.

_____   1

(4)

18. Ionic compounds such as potassium chloride can dissolve in water to form a solution.

(*a*) What **term** can be used to describe the water?

_____   1

(*b*) The graph shows how the temperature of the water affects the solubility of potassium chloride.

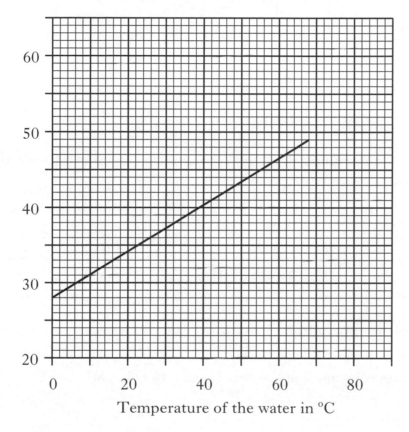

Solubility of potassium chloride in grams per $100\,cm^3$ of water

Temperature of the water in °C

(i) How does the temperature of the water affect the solubility of the potassium chloride?

_____

_____

_____   1

(ii) Predict the solubility of potassium chloride at 80 °C.

_____ grams per $100\,cm^3$ of water.   1

    (3)

[Turn over for Question 19 on *Page twenty-four*

Marks | KU | P:

19.   A student set up the cell shown.

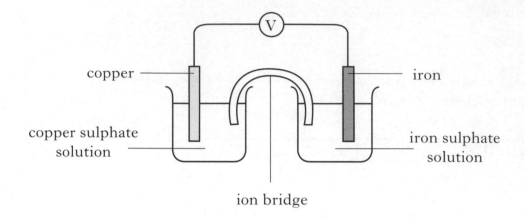

(*a*) **On the wires**, indicate the direction of electron flow.   **1**

(*b*) Why do ionic compounds, like copper sulphate, conduct electricity when in solution?

_____

_____

_____   **1**

(*c*) Name a metal which could be used to replace the iron to produce a smaller voltage.

You may wish to use page 7 of the data booklet to help you.

_____   **1**

(*d*) What is the purpose of the ion bridge?

_____

_____   **1**
                                                   (4)

[END OF QUESTION PAPER]

ADDITIONAL SPACE FOR ANSWERS

ADDITIONAL GRAPH PAPER FOR QUESTION 11(b)(i)

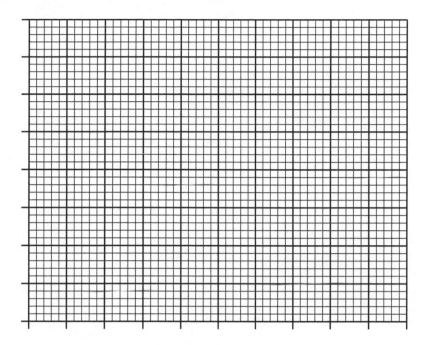

ADDITIONAL DIAGRAM FOR QUESTION 17(b)(i)

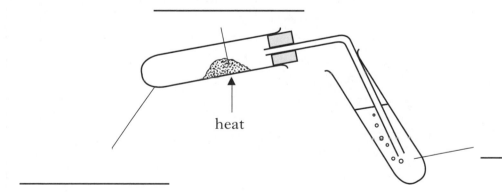

heat

ADDITIONAL SPACE FOR ANSWERS

STANDARD GRADE | GENERAL

2011

[BLANK PAGE]

FOR OFFICIAL USE

G

	KU	PS
Total Marks		

0500/401

NATIONAL QUALIFICATIONS 2011	THURSDAY, 26 MAY 9.00 AM – 10.30 AM	CHEMISTRY STANDARD GRADE General Level

---

Fill in these boxes and read what is printed below.

Full name of centre

Town

Forename(s)

Surname

Date of birth

Day	Month	Year	Scottish candidate number	Number of seat

1 All questions should be attempted.

2 Necessary data will be found in the Data Booklet provided for Chemistry at Standard Grade and Intermediate 2.

3 The questions may be answered in any order but all answers are to be written in this answer book, and must be written clearly and legibly in ink.

4 Rough work, if any should be necessary, as well as the fair copy, is to be written in this book.

Rough work should be scored through when the fair copy has been written.

5 Additional space for answers and rough work will be found at the end of the book.

6 The size of the space provided for an answer should not be taken as an indication of how much to write. It is not necessary to use all the space.

7 Before leaving the examination room you must give this book to the Invigilator. If you do not, you may lose all the marks for this paper.

PART 1

In Questions 1 to 8 of this part of the paper, an answer is given by circling the appropriate letter (or letters) in the answer grid provided.

In some questions, two letters are required for full marks.

If more than the correct number of answers is given, marks will be deducted.

A total of 20 marks is available in this part of the paper.

SAMPLE QUESTION

A CH_4	B H_2	C CO_2
D CO	E C_2H_5OH	F C

(a) Identify the hydrocarbon.

Ⓐ	B	C
D	E	F

The one correct answer to part (a) is A. This should be circled.

(b) Identify the **two** elements.

A	Ⓑ	C
D	E	Ⓕ

As indicated in this question, there are **two** correct answers to part (b). These are B and F. Both answers are circled.

If, after you have recorded your answer, you decide that you have made an error and wish to make a change, you should cancel the original answer and circle the answer you now consider to be correct. Thus, in part (a), if you want to change an answer A to an answer D, your answer sheet would look like this:

Ⓧ	B	C
Ⓓ	E	F

If you want to change back to an answer which has already been scored out, you should enter a tick (✓) in the box of the answer of your choice, thus:

✓Ⓐ	B	C
Ⓓ	E	F

DO NOT WRITE IN THIS MARGIN

Marks KU PS

1. The grid contains the symbols of some elements.

A Mg	B N	C Ag
D S	E F	F Si

(*a*) Identify the symbol for silver.

You may wish to use page 8 of the data booklet to help you.

A	B	C
D	E	F

1

(*b*) Identify the symbol for the element which has similar chemical properties to oxygen.

You may wish to use page 8 of the data booklet to help you.

A	B	C
D	E	F

1

(2)

[Turn over

Marks KU PS

2. A student set up four experiments to investigate the solubility of aspirin.

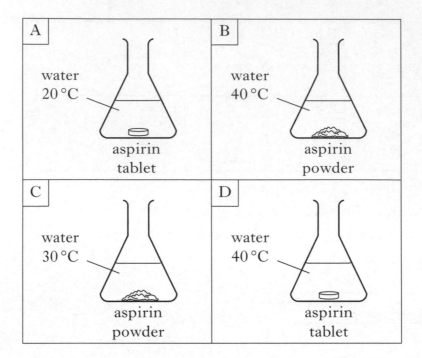

(*a*) Identify the experiment in which the aspirin would take the longest time to dissolve.

A	B
C	D

1

(*b*) Identify the **two** experiments which should be compared to show the effect of particle size on the speed of dissolving.

A	B
C	D

1

(2)

3. The grid shows the names of some metals.

A	B	C
sodium	calcium	potassium
D	E	F
zinc	tin	gold

(a) Identify the metal used to galvanise iron.

A	B	C
D	E	F

1

(b) Identify the metal that does **not** react with dilute acid.

A	B	C
D	E	F

1

(c) Identify the metal which has a relative atomic mass of 118·5.
You may wish to use page 4 of the data booklet to help you.

A	B	C
D	E	F

1

(d) Identify the metal found uncombined in the Earth's crust.

A	B	C
D	E	F

1

(4)

[Turn over

Marks | KU | PS

4. The grid contains the names of some chemical processes.

A electrolysis	B polymerisation	C Ostwald
D Haber	E distillation	F cracking

(a) Identify the process used to produce smaller, more useful hydrocarbons.

A	B	C
D	E	F

1

(b) Identify the process which uses electricity to break up a compound into its elements.

A	B	C
D	E	F

1

(c) Identify the process used to make plastics from alkenes.

A	B	C
D	E	F

1

(3)

5. The grid shows the formulae of some substances.

A He	B NO_2	C H_2
D K_2O	E O_2	F CO_2

(a) Identify the substance which is ionic.

A	B	C
D	E	F

1

(b) Identify the **two** substances which exist as **diatomic** molecules.

A	B	C
D	E	F

1

(c) Identify the substance produced in air during lightning storms.

A	B	C
D	E	F

1

(d) Identify the gas required for combustion to take place.

A	B	C
D	E	F

1

(4)

[Turn over

Marks KU P

6. The grid shows the names of some sulphates.

A ammonium sulphate	B sodium sulphate	C barium sulphate
D zinc sulphate	E copper sulphate	F magnesium sulphate

(*a*) Identify the sulphate which could be produced by a precipitation reaction.

You may wish to use page 5 of the data booklet to help you.

A	B	C
D	E	F

1

(*b*) Identify the sulphate which contains an essential element for healthy plant growth.

A	B	C
D	E	F

1

(2)

Marks | KU | PS

7. The burning of magnesium in nitrogen produces magnesium nitride.

When water is added to magnesium nitride, a gas is produced which turns pH paper blue.

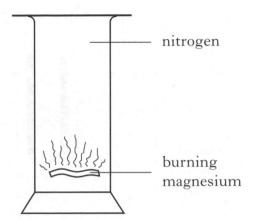

nitrogen

burning
magnesium

Identify the gas produced which turns pH paper blue.

A	oxygen
B	ammonia
C	hydrogen
D	carbon dioxide

A
B
C
D

(1)

[Turn over

Marks KU P

8. A teacher set up some experiments to investigate electrical conductivity.

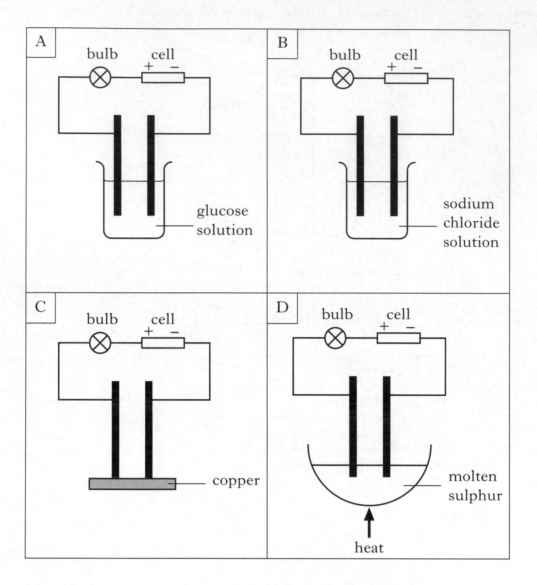

Identify the **two** experiments in which the bulb will light.

A	B
C	D

(2)

[Turn over for Part 2 on *Page twelve*

Marks | KU | P

PART 2

A total of 40 marks is available in this part of the paper.

9. The Periodic Table lists the names of elements.

   (*a*) Elements are made up of atoms.

   Why are atoms neutral?

   _____

   _____

   _____   1

   (*b*) The alkali metals, the halogens and the noble gases are groups of elements in the Periodic Table.

   Complete the table by (circling) a word in each box to give correct information about each group.

   (Two pieces of correct information have already been circled.)

   You may wish to use page 1 of the data booklet to help you.

Alkali Metals	Halogens	Noble Gases
(metal) / non-metal	metal / non-metal	metal / non-metal
reactive / unreactive	(reactive) / unreactive	reactive / unreactive

2

(3)

10. Hydrazine is a fuel used in rockets. The diagram represents a molecule of hydrazine.

represents a nitrogen atom (N)

represents a hydrogen atom (H)

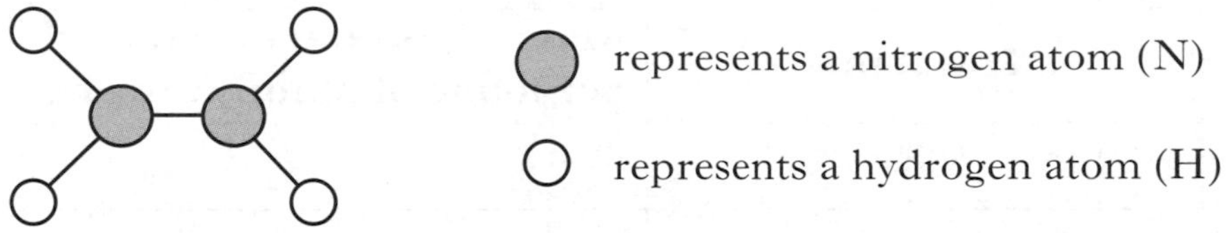

(a) Write the molecular formula for hydrazine.

(b) What holds the atoms together in a molecule of hydrazine?

_____

(c) Hydrazine is unstable and can break down to produce ammonia, nitrogen and hydrogen.

Write a **word** equation for this reaction.

[Turn over

Marks | KU | PS

11. (*a*) The table shows the mass of various pollutants produced by recycling aluminium.

Pollutant	Mass of pollutant produced per tonne of aluminium/kg
sulphur dioxide	1·0
dust	1·5
carbon monoxide	2·5
nitrogen oxides	7·0
hydrocarbons	5·0

Present the information as a bar chart.

Use appropriate scales to fill most of the graph paper.

(Additional graph paper, if required, can be found on page 28.)

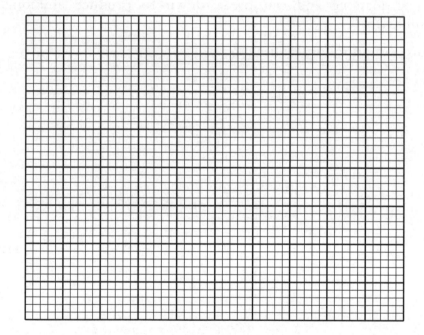

2

(*b*) When sulphur dioxide reacts with water in the atmosphere, acid rain is produced.

Give **one** example of a damaging effect of acid rain.

_____

_____

1

(3)

Marks | KU | PS

12. The experiment below was carried out to investigate what happens when different substances are heated and then allowed to cool.

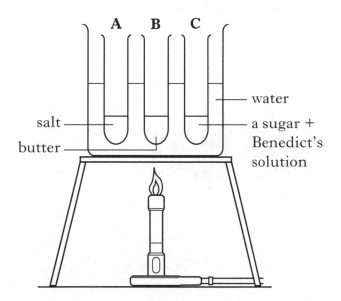

The results are shown.

Test tube	Contents	Observation on heating	Observation on cooling
A	salt	no change	no change
B	butter	yellow solid to yellow liquid	yellow liquid to yellow solid
C	a sugar and Benedict's solution	blue liquid to orange solid	orange solid remains

(a) In which test tube, **A**, **B** or **C**, did a chemical reaction take place?

_____

1

(b) Suggest a name for the sugar in test tube **C**.

_____

1

(2)

[Turn over

Marks KU PS

13. Forests are important in maintaining the level of carbon dioxide in the atmosphere.

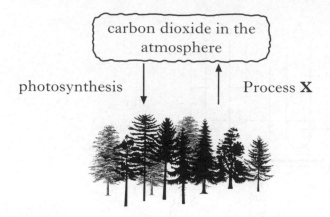

carbon dioxide in the atmosphere

photosynthesis   Process **X**

(a) Name Process **X**.

_____

1

(b) The equation for photosynthesis is:

carbon dioxide + compound **Y** ⟶ glucose + oxygen

Name compound **Y**.

_____

1

(c) The table shows how the level of carbon dioxide in the atmosphere has changed since 1975.

Year	Level of carbon dioxide/units
1975	330
1985	345
1995	358
2005	374
2015	

Predict the level of carbon dioxide in the atmosphere in 2015 if the trend continues.

_____ units

1

(3)

14. (*a*) The use of a metal depends on its properties.

Complete the table to show the property of the metal which makes it suitable for each use.

Choose your answer from the following.

conducts heat   low density   conducts electricity

Use of metal	Property of metal
aeroplane	
overhead cables	
cooking pot	

(*b*) Most metals in the Earth's crust are found combined with other elements.

What **term** is used to describe naturally occurring compounds of metals?

_____

1

(2)

[Turn over

Marks | KU | PS

15. Some supermarkets no longer supply free polythene bags because they are non-biodegradable and can cause environmental problems.

(*a*) (i) What does non-biodegradable mean?

_____

_____   1

(ii) Draw a section of polythene, showing 3 monomer units joined together.

$$\begin{array}{cccccc} \text{H} & \text{H} & \text{H} & \text{H} & \text{H} & \text{H} \\ | & | & | & | & | & | \\ \text{C} = \text{C} & + & \text{C} = \text{C} & + & \text{C} = \text{C} \\ | & | & | & | & | & | \\ \text{H} & \text{H} & \text{H} & \text{H} & \text{H} & \text{H} \end{array}$$

↓

1

15. (continued)

(*b*) The graph shows the mass of polythene used in a European country.

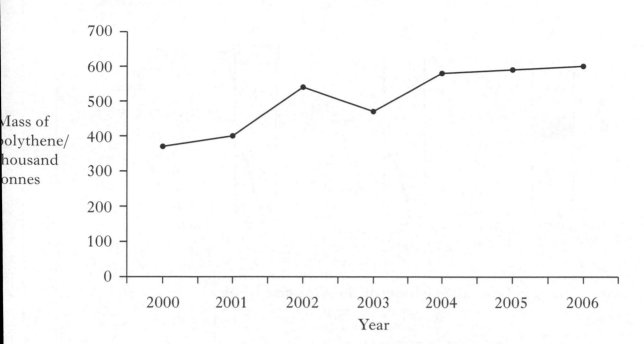

Mass of polythene/ thousand tonnes

Year

Describe the **general** trend in the mass of polythene used between 2000 and 2006.

_____

1

(3)

[Turn over

Marks | KU | P

16. A student investigated the rusting of iron.

(*a*) He set up three test tubes each containing a clean iron nail.

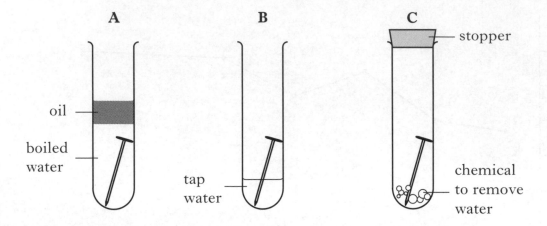

Test tube	Observation after one week
A	Nail stayed bright
B	Nail rusted
C	Nail stayed bright

Suggest why the nail in test tube **A** did not rust.

_____

_____

1

16.   (continued)

(b)   The student also set up two dishes containing clean iron nails set in a gel containing ferroxyl indicator.

The diagram below shows the result after 1 day.

iron nail

blue
colour

magnesium
connected to
an iron nail

(i)   Write the symbol for the iron ion which turns ferroxyl indicator blue.

_____

1

(ii)   Explain **why** the magnesium connected to the iron nail prevents rusting.

_____

_____

1

(3)

[Turn over

Marks KU PS

17.   **Manufacture of Malt Vinegar**

In the manufacture of malt vinegar, starch reacts with water to produce glucose.

Glucose is converted to carbon dioxide and ethanol using an enzyme.

Ethanol is oxidised, by bacteria, producing malt vinegar.

(*a*)   Use the information to complete the flow diagram.

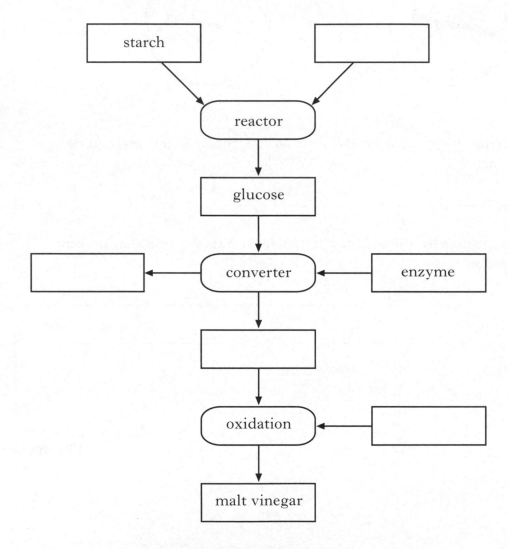

2

17. (continued)

(b) A student measured the pH of malt vinegar and other substances.

The results are shown.

Substance	pH value
malt vinegar	
oven cleaner	11
water	7
lemon juice	5
liquid soap	8

(i) Describe how the student would have used universal indicator or pH paper to measure the pH values.

_____

_____

_____ **1**

(ii) Malt vinegar contains ethanoic acid.

Suggest a pH value for malt vinegar.

_____ **1**

(c) Lemon juice contains citric acid.

Complete the sentence below by (circling) the correct answer.

When lemon juice is diluted with water the pH $\left\{\begin{array}{l}\text{decreases}\\\text{stays the same}\\\text{increases}\end{array}\right\}$. **1**

(5)

[Turn over

Marks | KU | PS

18. Crude oil is a fossil fuel.

(*a*) Name another fossil fuel.

_____

1

(*b*) Crude oil can be separated into fractions.

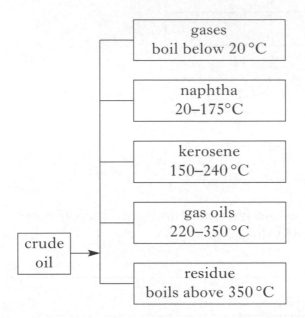

(i) Identify the fraction in which butane is present.

You may wish to use page 6 of the data booklet to help you.

_____

1

(ii) The table shows information about the colour of each fraction

Fraction	Colour
gases	colourless
naphtha	light yellow
kerosene	dark yellow
gas oils	brown
residue	black

What is the colour of the fraction which is collected at 250 °C?

_____

1

(3)

Marks KU PS

19. A student investigated the effect of concentration on the rate of reaction between magnesium and sulphuric acid.

In each case she used the same mass of magnesium ribbon and timed how long it took for the magnesium to disappear.

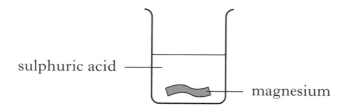

The results are shown.

	Volume of 2 mol/l sulphuric acid/cm³	Volume of water/cm³	Total volume/cm³	Time/s
Experiment 1	20	0	20	50
Experiment 2	15		20	65

(a) (i) Complete the table to show the volume of water the student should have used in experiment **2**.

1

(ii) How did the **speed** of the reaction in experiment **2** compare with the speed of the reaction in experiment **1**?

_____

1

(b) Magnesium reacts with dilute sulphuric acid to produce magnesium sulphate and hydrogen gas.

State the test for hydrogen gas.

_____

_____

1

(3)

[Turn over

Marks   KU   P

20. The key below shows the name and general formula of some families of hydrocarbons.

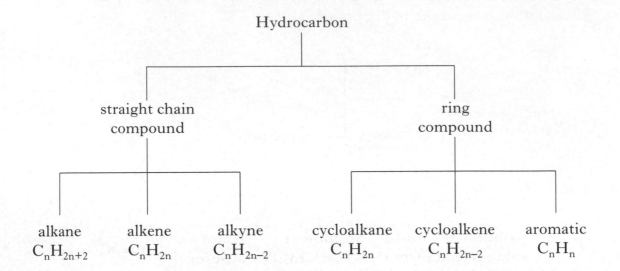

(a) (i) Butyne is an alkyne with 4 carbon atoms.

Using the key, write the **molecular** formula for butyne.

1

(ii) Hydrocarbon **X** is a ring compound with molecular formula C_6H_6.

Using the key, name the **family** to which it belongs.

_____

1

(b) Draw a structural formula for the alkane pentane.

1

(c) Describe the chemical test, including the result, which shows that an alkene is unsaturated.

_____

_____

1

(4)

OFFICIAL SQA PAST PAPERS   119   GENERAL CHEMISTRY 2011

DO NOT
WRITE IN
THIS
MARGIN

Marks | KU | PS

21. (a) A student set up the following cell.

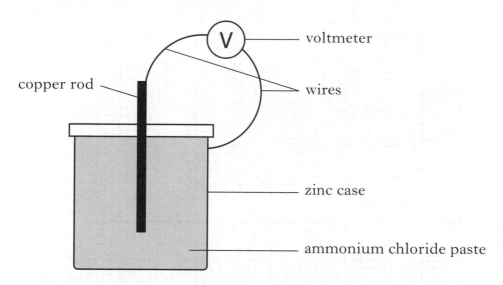

voltmeter

copper rod

wires

zinc case

ammonium chloride paste

(i) In this cell, the purpose of the ammonium chloride is to complete the circuit.

What **term** is used to describe an ionic compound, like ammonium chloride, which is used for this purpose?

_____

1

(ii) (Circle) the correct word, in each bracket, to complete the sentence below.

You may wish to use page 7 of the data booklet to help you.

In the cell the electrons flow from $\begin{Bmatrix} copper \\ zinc \end{Bmatrix}$ to $\begin{Bmatrix} copper \\ zinc \end{Bmatrix}$ through the $\begin{Bmatrix} paste \\ wires \end{Bmatrix}$.

1

(b) A battery is a number of cells joined together.

Why do batteries stop producing electricity after some time?

_____

_____

1

(3)

[END OF QUESTION PAPER]

ADDITIONAL SPACE FOR ANSWERS

ADDITIONAL GRAPH PAPER FOR QUESTION 11(*a*)

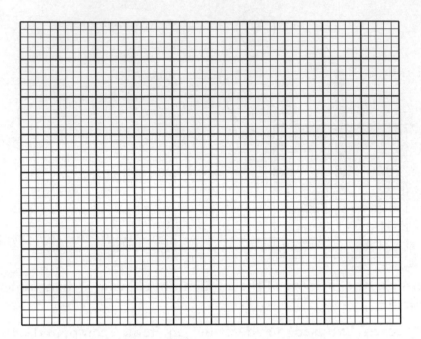

ADDITIONAL SPACE FOR ANSWERS

ADDITIONAL SPACE FOR ANSWERS

ADDITIONAL SPACE FOR ANSWERS

ADDITIONAL SPACE FOR ANSWERS

DO NOT
WRITE IN
THIS
MARGIN

KU PS

[BLANK PAGE]

FOR OFFICIAL USE

KU   PS

Total
Marks

G

0500/29/01

NATIONAL
QUALIFICATIONS
2012

MONDAY, 14 MAY
9.00 AM – 10.30 AM

CHEMISTRY
STANDARD GRADE
General Level

Fill in these boxes and read what is printed below.

Full name of centre

Town

Forename(s)

Surname

Date of birth

Day	Month	Year	Scottish candidate number	Number of seat

1   All questions should be attempted.

2   Necessary data will be found in the Data Booklet provided for Chemistry at Standard Grade and Intermediate 2.

3   The questions may be answered in any order but all answers are to be written in this answer book, and must be written clearly and legibly in ink.

4   Rough work, if any should be necessary, as well as the fair copy, is to be written in this book.
    Rough work should be scored through when the fair copy has been written.

5   Additional space for answers and rough work will be found at the end of the book.

6   The size of the space provided for an answer should not be taken as an indication of how much to write.   It is not necessary to use all the space.

7   Before leaving the examination room you must give this book to the Invigilator. If you do not, you may lose all the marks for this paper.

PART 1

In Questions 1 to 9 of this part of the paper, an answer is given by circling the appropriate letter (or letters) in the answer grid provided.

In some questions, two letters are required for full marks.

If more than the correct number of answers is given, marks will be deducted.

A total of 20 marks is available in this part of the paper.

SAMPLE QUESTION

A CH_4	B H_2	C CO_2
D CO	E C_2H_5OH	F C

(a)   Identify the hydrocarbon.

Ⓐ	B	C
D	E	F

The one correct answer to part (a) is A. This should be circled.

(b)   Identify the **two** elements.

A	Ⓑ	C
D	E	Ⓕ

As indicated in this question, there are **two** correct answers to part (b). These are B and F.

Both answers are circled.

If, after you have recorded your answer, you decide that you have made an error and wish to make a change, you should cancel the original answer and circle the answer you now consider to be correct. Thus, in part (a), if you want to change an answer A to an answer D, your answer sheet would look like this:

Ⱥ	B	C
Ⓓ	E	F

If you want to change back to an answer which has already been scored out, you should enter a tick (✓) in the box of the answer of your choice, thus:

✓Ⱥ	B	C
Ⱦ	E	F

1. The grid shows the names of some elements.

A gold	B magnesium	C carbon
D nitrogen	E calcium	F iodine

(a) Identify the element with atomic number 79.

You may wish to use page 8 of the data booklet to help you.

A	B	C
D	E	F

1

(b) Identify the **two** elements which exist as diatomic molecules.

A	B	C
D	E	F

1

(c) Identify the **two** elements which have similar chemical properties.

You may wish to use page 8 of the data booklet to help you.

A	B	C
D	E	F

1

(3)

[Turn over

Marks KU P

2. A catalyst speeds up the following reaction:

hydrogen peroxide ⟶ water + oxygen

The grid shows reactions carried out using the **same** mass of catalyst with two different concentrations of hydrogen peroxide.

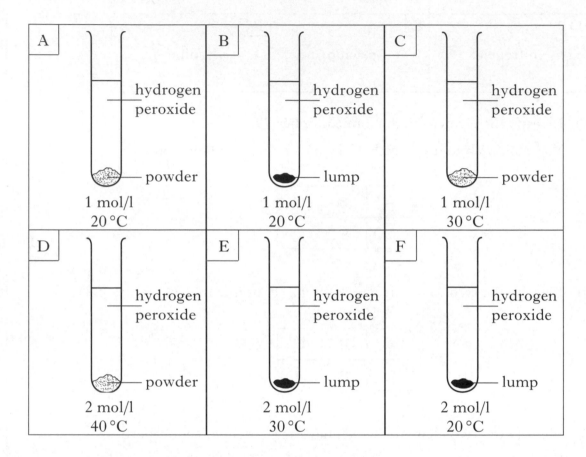

(*a*) Identify the **two** experiments which could be used to show the effect of concentration on the speed of reaction.

A	B	C
D	E	F

1

(*b*) Identify the experiment with the fastest speed of reaction.

A	B	C
D	E	F

1

(2)

DO NOT WRITE IN THIS MARGIN

Marks KU PS

3. The grid shows the names of some substances.

A potassium	B water	C helium
D air	E sodium chloride	F phosphorus

(*a*) Identify the **two** non-metal elements.

You may wish to use page 1 of the data booklet to help you.

A	B	C
D	E	F

1

(*b*) Identify the mixture.

A	B	C
D	E	F

1

(2)

[Turn over

Marks KU PS

4. The grid shows the names of some metals.

A silver	B sodium	C magnesium
D nickel	E lead	F iron

(*a*) Identify the metal produced in a Blast Furnace.

A	B	C
D	E	F

1

(*b*) Identify the metal that does **not** react with dilute acid.

You may wish to use page 7 of the data booklet to help you.

A	B	C
D	E	F

1

(*c*) Identify the metal that is stored under oil.

You may wish to use page 8 of the data booklet to help you.

A	B	C
D	E	F

1

(3)

DO NOT WRITE IN THIS MARGIN

Marks KU PS

5.

A butter melting	B distillation of crude oil
C wood burning	D water evaporating

Identify the chemical reaction.

A	B
C	D

(1)

[Turn over

DO NOT WRITE IN THIS MARGIN

Marks KU PS

6. The grid shows the names of some compounds.

A zinc chloride	B magnesium sulphite	C sodium chlorate
D lead carbonate	E hydrogen sulphide	F potassium nitrite

(*a*) Identify the **two** compounds which do not contain oxygen.

A	B	C
D	E	F

1

(*b*) Identify the covalent compound.

A	B	C
D	E	F

1

(2)

DO NOT WRITE IN THIS MARGIN

Marks | KU | PS

7. The grid shows the names of some gases.

A chlorine	B nitrogen	C ammonia
D oxygen	E hydrogen	F ethene

(a) Identify the gas which is a hydrocarbon.

A	B	C
D	E	F

1

(b) Identify the gas which turns damp pH paper blue.

A	B	C
D	E	F

1

(c) Identify the gas produced when dilute hydrochloric acid reacts with zinc.

A	B	C
D	E	F

1

(3)

[Turn over

DO NOT WRITE IN THIS MARGIN

Marks KU PS

8. The grid shows the formulae of some ions.

A H^{+}	B NO_3^{-}	C Fe^{2+}
D OH^{-}	E SO_4^{2-}	F Na^{+}

(*a*) Identify the ion which turns ferroxyl indicator blue.

A	B	C
D	E	F

1

(*b*) Identify the ion that can be used as a fertiliser.

A	B	C
D	E	F

1

(2)

Marks

9. A student made some statements about the rusting of iron.

A	Both air and water are required for rusting to occur.
B	Attaching iron to the positive terminal of a battery prevents rusting.
C	Salt slows down rusting.
D	Rusting is the corrosion of iron.
E	Coating iron in nickel is called galvanising.

Identify the **two** correct statements.

A
B
C
D
E

(2)

[Turn over for Part 2 on *Page twelve*

Marks KU PS

PART 2

A total of 40 marks is available in this part of the paper.

10. An atom of fluorine can be represented by a simple diagram.

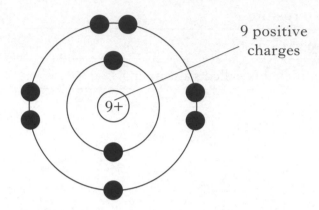

9 positive
charges

(*a*) Name the structure at the centre of the atom where the positive charges
are found.

_____   1

(*b*) Fluorine is found in group 7 of the Periodic Table.

Name the family of elements to which fluorine belongs.

_____   1

(2)

DO NOT
WRITE
IN THIS
MARGIN

Marks | KU | PS

11. Metals can be used as catalysts.

Catalyst	Use
platinum	catalytic converter
nickel	making margarine
iron	making ammonia
rhodium	drug manufacture

(*a*) Name the industrial process used to make ammonia.

_____

1

(*b*) Platinum is found uncombined in the Earth's crust.

What does this indicate about the reactivity of platinum?

_____

1

(*c*) Metals can also be used to make alloys.

Name an alloy.

_____

1

(3)

[Turn over

Marks

12. One way of classifying the types of hydrocarbon found in crude oil is shown in the table.

Type of hydrocarbon	% in crude oil
naphthalenes	50
paraffins	30
aromatics	15
asphalts	

(a) Label the pie chart to show the name and percentage for each type of hydrocarbon.

One label has already been completed for you.

(An additional pie chart, if required, can be found on page 27.)

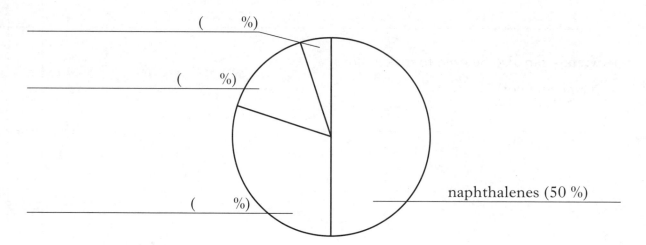

(%)

(%)

(%)

naphthalenes (50 %)

2

DO NOT WRITE IN THIS MARGIN

Marks KU PS

12. (continued)

(*b*) The table below gives information about some hydrocarbons obtained from the paraffins.

Name	Formula
octane	C_8H_{18}
nonane	C_9H_{20}
decane	$C_{10}H_{22}$
undecane	$C_{11}H_{24}$

Name the family of hydrocarbons in the table.

________________________________ **1**

(*c*) Eicosane is another member of this family.

A molecule of eicosane contains 20 carbon atoms.

Write the molecular formula of eicosane.

1

(4)

[Turn over

Marks | KU | PS

13. Vinegar is a solution of ethanoic acid in water.

(*a*) A student set up the following experiments.

He tested ethanoic acid solutions of different concentrations.

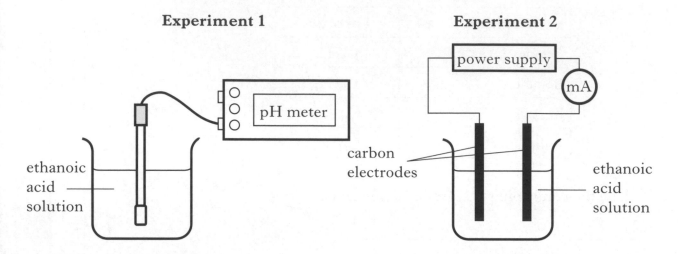

Experiment 1   **Experiment 2**

His results are shown below.

Ethanoic acid solution	pH	Current/mA
A	3	18
B	4	9
C	5	5

(i) Which ethanoic acid solution is the most acidic, **A**, **B** or **C**?

_____   1

(ii) Predict the current, in mA, for an ethanoic acid solution of pH 6.

_____ mA   1

(*b*) Name the ion present in all acidic solutions.

_____   1

                                                             (3)

Marks　KU　PS

14. Polystyrene is a plastic used in packaging.

(a) Name the monomer used to make polystyrene.

_____

1

(b) Name the type of chemical reaction which is used to make polystyrene.

_____

1

(c) Starch, obtained from natural sources such as barley, can be used to make a packaging material with similar properties to polystyrene.

Suggest one advantage of this material compared to polystyrene.

_____

_____

1

(3)

[Turn over

Marks KU PS

15. (*a*) The carbohydrate glucose is made when green plants absorb light energy from the sun.

sunlight

(i) Name the chemical, present in green plants, which absorbs light energy.

_____   **1**

(ii) Describe the chemical test, including the result, for glucose.

_____

_____   **1**

Marks | KU | PS

15. **(continued)**

(b) A student set up an experiment to investigate the burning of carbohydrates.

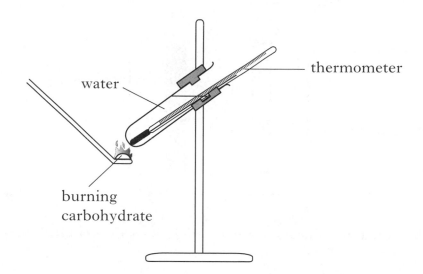

water

thermometer

burning
carbohydrate

Her results are shown below.

Carbohydrate	Starting temperature of water/° C	Final temperature of water/° C
glucose	20	44
starch	20	56

Suggest **one** factor that the student would have kept the same to make a fair comparison.

_____ 1

(c) Circle the correct words to complete the sentence.

Starch is $\begin{Bmatrix} \text{sweet} \\ \text{not sweet} \end{Bmatrix}$ and $\begin{Bmatrix} \text{dissolves} \\ \text{does not dissolve} \end{Bmatrix}$ well in water.   1

(d) Scientists have developed a method of producing hydrocarbons from carbohydrates.

Name the element removed from a carbohydrate to produce a hydrocarbon.

_____ 1

(5)

Marks   KU   PS

16. The diagram below shows a cell.

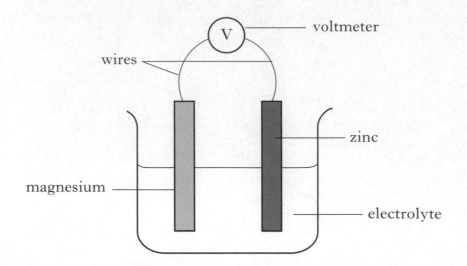

(a) Name the type of charged particle that flows through the wires.

_____   1

(b) The voltage of the cell shown above is 1·51 V.

Name a metal which could replace **zinc** to produce a **greater** voltage.

You may wish to use page 7 of the data booklet to help you.

_____   1

(c) Scientists at the University of St. Andrews have developed a type of battery. It has the advantage of being able to store up to 10 times more energy than some other types of battery.

   (i) Suggest another advantage of using this type of battery.

_____

_____   1

  (ii) The chemical reaction inside this battery produces lithium oxide.

Write the formula for lithium oxide.

1

16. (continued)

(*d*) The table below shows the maximum storage life of some other types of battery.

Type of battery	Storage life/years
alkaline	5
zinc chloride	2
silver oxide	2
nickel-cadmium	7
lithium	10

Present the information as a bar chart.

Use appropriate scales to fill most of the graph paper.

(Additional graph paper, if required, can be found on page 27.)

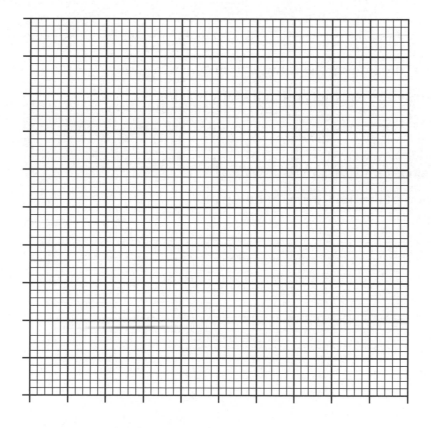

2

(6)

[Turn over

DO NOT
WRITE
IN THIS
MARGIN

Marks KU P

17. A teacher demonstrated an experiment to show how nitrogen dioxide is formed in a petrol engine.

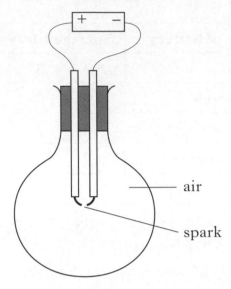

high voltage supply

air

spark

(*a*) Name the **two** gases which react to form nitrogen dioxide.

_____ 1

(*b*) Nitrogen dioxide can be formed naturally in air.

What provides the high voltage spark for this reaction?

_____ 1

(*c*) Nitrogen dioxide dissolves in water.

Suggest a pH value for this solution.

_____ 1

(3)

18. The diagram shows an arrangement of ions in an ionic compound.

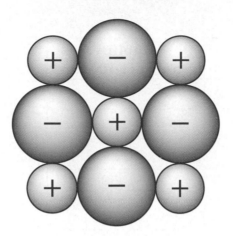

(a) What term is given to the arrangement of ions in an ionic solid?

_____   1

(b) Explain why solid ionic compounds do **not** conduct electricity.

_____

_____   1

(c) Many ionic compounds are coloured.

Compound	Colour
copper sulphate	blue
nickel chloride	green
sodium dichromate	orange
sodium chloride	colourless

Using the information in the table, state the colour of the chloride ion.

_____   1

(d) Copper can be extracted from the ionic compound copper oxide as shown.

copper oxide + **Y** ⟶ copper + carbon dioxide

Name **Y**.

_____   1

(4)

DO NOT WRITE IN THIS MARGIN

Marks KU PS

19. (*a*) The table gives information on the solubility of some compounds in water.

Compound	**Solubility/ grams per 100 cm^3**
potassium chlorate	10·0
potassium nitrate	33·4
sodium carbonate	7·1
sodium chloride	36·5
sodium nitrate	88·6

Using the information in the table, name the **least** soluble compound.

______________________________ 1

OFFICIAL SQA PAST PAPERS   151   GENERAL CHEMISTRY 2012

DO NOT
WRITE
IN THIS
MARGIN

Marks KU PS

19. (continued)

(*b*) The graph shows the solubility of sodium chloride and potassium nitrate at different temperatures.

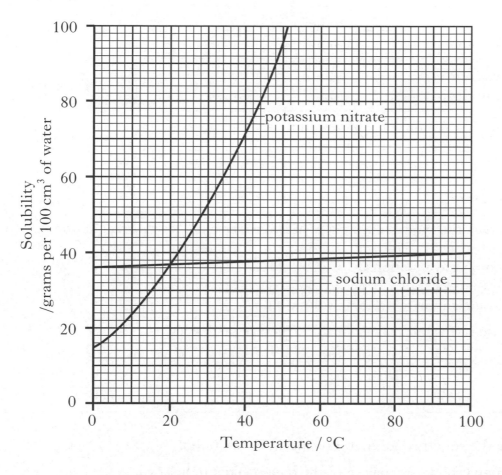

(i) At what temperature do sodium chloride and potassium nitrate have the **same** solubility?

_____ °C   1

(ii) Write a general statement describing the effect of temperature on the solubility of potassium nitrate.

_____

_____   1

(3)

[Turn over for Question 20 on *Page twenty-six*

Marks

20. The table shows word equations for some chemical reactions.

	Word Equation					Type of chemical reaction
A	large alkane	→	smaller alkane	+	alkene	_____
B	lead nitrate + sodium iodide	→	sodium nitrate	+	lead iodide	precipitation
C	potassium hydroxide + hydrochloric acid	→	potassium chloride	+	_____	neutralisation

(a) **In the table,**

    (i)   write the type of chemical reaction represented by word equation **A**;   1

    (ii) complete equation **C**.   1

(b) Alkenes decolourise bromine solution.

What does this tell you about the structure of alkenes?

_____

_____   1

(c) Name the solid produced in precipitation reaction **B**.

You may wish to use page 5 of the data booklet to help you.

_____   1

(4)

[*END OF QUESTION PAPER*]

OFFICIAL SQA PAST PAPERS   153   GENERAL CHEMISTRY 2012

DO NOT
WRITE
IN THIS
MARGIN

KU | PS

ADDITIONAL SPACE FOR ANSWERS

ADDITIONAL PIE CHART FOR QUESTION 12(a)

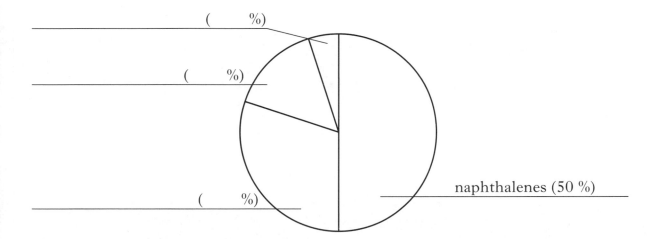

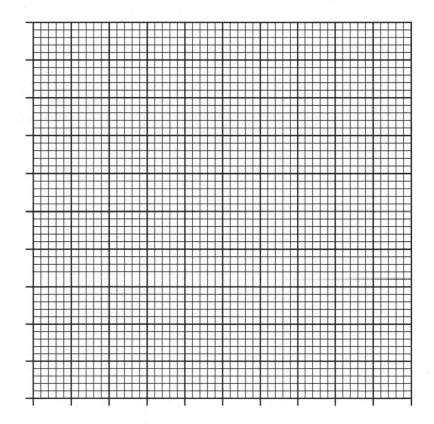

ADDITIONAL GRAPH PAPER FOR QUESTION 16(d)

ADDITIONAL SPACE FOR ANSWERS

STANDARD GRADE | ANSWER SECTION

SQA STANDARD GRADE
GENERAL CHEMISTRY 2008–2012

CHEMISTRY GENERAL 2008

PART 1

1. (a) B and D

   (b) D

   (c) F

   (d) A and E

2. (a) A and C

   (b) E

   (c) C

3. (a) C

   (b) B

4. (a) C and E

   (b) E

   (c) A

5. (a) D

   (b) A

6. D

7. (a) E

   (b) F

8. B

9. B and D

PART 2

10. (a) Distillation/fractional distillation

    (b) Smaller

    (c) Aluminum (Al), silicon (Si), oxygen (O)

11. (a) Burns with a pop/lit splint and it pops

    (b) A solution which turns universal indicator purple is _Alkaline_.

    (c) _Any one from:_
    • Too reactive/react violently/very reactive
    or
    • Prevents it reacting with air/oxygen/water
    or
    • Prevents oxidation/corrosion

12. (a)

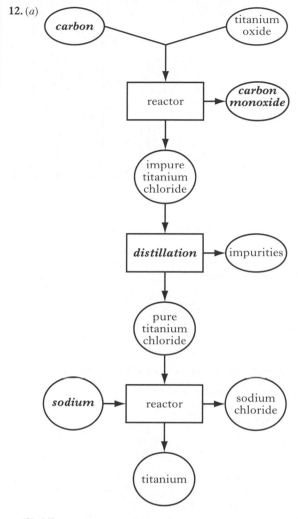

    (b) Alloy

    (c)

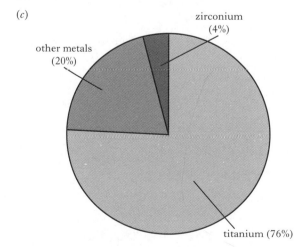

13. (a) (i) Glowed very brightly/brighter than zinc/burns/bright flame/white light
        (ii) Faster/increase

    (b) Gold / mercury / platinum

14. (a) Carbohydrates/saccharides

    (b) (i) Fermentation/anaerobic respiration
        (ii) Ethanol

15. (*a*)

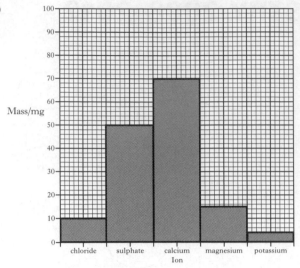

(*b*) (i) CO_2

(ii)   It would turn milky/chalky/cloudy white

16. (*a*) Parsnip

(*b*) Enzyme

(*c*) 20 cm³
25°C
1 g

17. (*a*) Chloroethene

(*b*) Last long(er) or will not rot/decay/decompose/break down

(*c*) (i)   Stronger/greater/higher
Natural are weaker

(ii)   Natural

18. (*a*) Hydrogen sulphide + oxygen → sulphur dioxide + water

(*b*) (i)   To speed up reaction / less heat / energy
(ii)   It decreases/goes down/gets lower or decreasing temperature increases %

19. (*a*) Acid

(*b*) 2

(*c*) A substance which burns/reacts with oxygen releasing energy/heat

(*d*) Carbon dioxide (CO_2)
Water (H_2O)

20. (*a*) Lettuce

(*b*) Lime/carbonate/base/an alkali/alkali fertiliser/any named alkali/carbonate/base

21. (*a*) Any number lower than 7

(*b*) More food needed/larger population so more food required

(*c*) Zinc chloride ($ZnCl_2$)

CHEMISTRY GENERAL 2009

PART 1

1. (*a*) B   (*b*) E   (*c*) C and D
2. (*a*) D   (*b*) A   (*c*) F
3. (*a*) B and F   (*b*) E
4. (*a*) C   (*b*) B   (*c*) E   (*d*) B
5. (*a*) F   (*b*) D
6. (*a*) B   (*b*) C
7. A and C
8. A and E

PART 2

9. (*a*) Exothermic

(*b*) Will run out/will not last forever
limited amount
non-renewable

(*c*) Oil/gas/peat
Natural gas/crude oil

10. (*a*) Magnesium   12   metal
Chlorine   17   non-metal

(*b*) (i) $MgCl_2$

(ii)   Boiling point = 1418°C
Melting point = 712°C

(iii)   Liquid

11. (*a*) Table drawn
Suitable headings
Correct entries

(*b*) Rot/rot away/broken down/disintegrate/deteriorate

(*c*) Thermoplastic

(*d*) Styrene

(*e*) Addition polymerisation/polymerisation

12. (*a*) Chlorine/chlorine gas/Cl_2
(*b*) Electrons/e^-
(*c*) Carbon/graphite/C

13. (*a*) 0·8g/the same

(*b*) (i)   Cracking

(ii) C_7H_{14}

14. (*a*) Vertical scale and label
Correct bar labelling
Bars drawn correctly

(*b*) Two atoms/two atoms joined/two oxygen atoms/ atoms go around in pairs

(*c*) Alloy

(*d*) (i)   Stops either or both air/oxygen/water/moisture

(ii)   Magnesium/zinc/aluminium (or correct symbol)

15. (*a*) Run out/expensive
Chemicals run out/used up
Limited supply of energy/power

(b) (i) Voltage higher than 0·5 but lower than 2·7

    (ii) Volume/type/concentration of electrolyte/solution
       Depth of immersion of rods
       Size of rods (electrodes)
       Separation of rods (electrodes)
       Temperature

16. *(a)* (i) Man-made/not natural

    (ii) Phosphorus or P   Potassium or K

  (b) Ammonia/NH_3

17. *(a)* Biological catalyst

  (b) $C_3H_6O_3$ (symbols in any order)

  (c) So that alcohol is not produced
     So the sugar would not ferment
     Sugar would be used up/react/break down

18. *(a)* Compare/check result with colour (pH) chart

  (b) Any number less than 7
     (0 – 6·9 inclusive)

  (c) Aluminium oxide is insoluble

19. *(a)* The higher the number of carbon atoms the lower the
     octane number or the lower the number of carbon atoms
     the higher the octane number
     The octane number decreases as the number of carbon
     atoms increases or the octane number increases as the
     number of carbon atoms decreases

  (b) Greater than 98 but less than 109
     (99 – 108 inclusive)

  (c) Alkane octane number is lower/it is lower
     It is less than the octane number of the alkenes
     Alkene number is higher
     Alkene is higher/alkane is lower

CHEMISTRY GENERAL 2010

PART 1

1. *(a)* B   *(b)* E   *(c)* F
2. *(a)* C   *(b)* E
3. *(a)* D and F   *(b)* A and E
4. *(a)* B   *(b)* A
5. *(a)* D and F   *(b)* C
6. *(a)* B   *(b)* E   *(c)* C
7. *(a)* A   *(b)* C and E
8. A and E
9. A and D

PART 2

10. *(a)* Millions
     Sea bed   (both required)

  (b) Sulphur dioxide
     SO_2

  (c) Bonds

     Shared pair of electrons (shared electrons)
     Attraction between shared electrons and nuclei/both nucleus

11. *(a)* (i) Re-lights a glowing splint
       Glowing splint bursts into flames/catches fire
    (ii) X = carbon dioxide/CO_2
       Y = water//H_2O/moisture

  (b) (i)

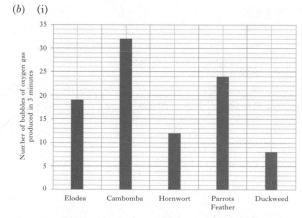

     Vertical scale + label (½ mark)
     Correct bar labelling (½ mark)
     Bars drawn correctly (1 mark)

    (ii) Temperature of water/heat from lamp
       Distance of lamp from plant/distance of lamp/same
       lamp/bulb
       Volume of water
       Brightness/power/wattage
       Light intensity/amount of light/colour of light
       Amount/size of plant/mass/surface area
       Same size of beaker/test tube/weight of plant

12. *(a)* C_4H_{10} + O_2 → CO_2 + H_2O

  (b) (i) Conducts heat
    (ii) Thermosetting
       Thermoset

  (c) Carbon/soot/C

13. (a) Potassium   or   K
Carbon   or   C
Oxygen   or O/O_2
(all three required)

   (b) (i) Precipitation
       (ii) Potassium nitrate
           or
           KNO_3
      (iii) Filtration/correct description of filtration/filter

14. (a)

Element mixed with iron	Use
chromium	cooking pots
manganese	railway tracks
titanium	aircraft parts
tungsten	hammers

    Table drawn ($\frac{1}{2}$ mark)
    Suitable headings ($\frac{1}{2}$ mark)
    Correct entries (1 mark)

   (b) (i) Water (H_2O)
          Moisture   } ($\frac{1}{2}$ mark)
          Rain

          Oxygen (O or O_2) } ($\frac{1}{2}$ mark)
          Air

      (ii) Sacrificial

   (c) Paint/attach negative terminal of battery/greasing/oiling
galvanising/coat it in plastic/tin plate/zinc coat/
electroplating/cathodic/metal coating

15. (a) Diatomic

   (b) Electrolysis

   (c) (i) Less CO_2 produced/lower carbon emissions
        Water is produced/no carbon dioxide produced
        No/less harmful/greenhouse/poisonous gases
        Less harmful to environment
        Renewable
        Not using 'finite' resources
        No/less pollution
      (ii) Air/atmosphere/Water
     (iii) Speed up reaction
         Lower temperature/less energy

16. (a) Potassium (K)
     or
     Phosphorus (P)

   (b)   Insoluble
       Does not dissolve

   (c)   Roots/nodules

17. (a) Alkanes

   (b) (i) Octane/boiling tube/
        or test tube
        Aluminium oxide/catalyst
        Bromine solution/water
      (ii) C_5H_{12}
        Pentane
        Correct structural formula
     (iii) Poly(propene)
        Polypropene
        Polypropylene

18. (a) Solvent

   (b) (i) As the temperature of the water { increases / decreases

        the solubility { increases / decreases .
        or
        The solubility { increases / decreases

        as the temperature { increases / decreases .

        As the temperature of water increases the solubility
does as well

     (ii) 53 (+/−1)

19. (a) Arrow drawn on diagram from iron to copper/right to left

   (b) Ions able to move (in solution)
      Free ions

   (c) Nickel   (Ni)
      Tin   (Sn)
      Lead   (Pb)
      Silver   (Ag)
      Mercury (Hg)

   (d) To complete/finish the circuit/cell
      Allow ions to move/flow/transfer ions

CHEMISTRY GENERAL 2011

PART 1

1. (a) C   (b) D
2. (a) A   (b) B and D
3. (a) D   (b) F   (c) E   (d) F
4. (a) F   (b) A   (c) B
5. (a) D   (b) C and E   (c) B   (d) E
6. (a) C   (b) A
7. B
8. B and C

PART 2

9. (a) Same or equal number of protons and electrons
   or
   Same or equal number of positive and negative charges

   (b) Alkali Metals   reactive

   Halogens   non-metal

   Noble Gases   non-metal   unreactive

10. (a) N_2H_4 or H_4N_2

    (b) Bond/covalent bond
    Shared pair of electrons
    Attraction between shared electrons and (positive) nuclei

    (c) Hydrazine $\longrightarrow$ ammonia + nitrogen + hydrogen

11. (a)

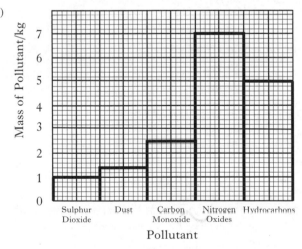

Vertical scale + label including unit – $\frac{1}{2}$ mark
Correct bar labelling – $\frac{1}{2}$ mark
Bars drawn correctly – 1 mark

(b) Kills/destroys/harms any named living thing
or
erodes/destroys or wears away stone/rocks/buildings

12. (a) C

    (b) Glucose/fructose/maltose or any other named reducing sugar

13. (a) (aerobic) respiration

    (b) Water/H_2O/hydrogen oxide

    (c) 383–394 inclusive

14. (a) low density
    conducts electricity
    conducts heat

    (b) (metal) ore

15. (a) (i) Will not rot/decompose/decay/cannot be broken down by bacteria

    (ii)

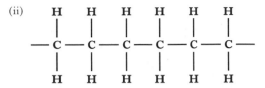

    (b) Increase/goes up/upward trend

16. (a) no air/oxygen

    (b) (i) Fe^{2+}/Fe^{+2}

    (ii) magnesium higher in electrochemical series
    or
    magnesium more reactive
    or
    magnesium provides sacrificial protection

17. (a) water
    carbon dioxide
    ethanol
    bacteria

    (b) (i) Put pH paper or universal indicator into substance
    Match to pH/colour chart/pH scale

    (ii) Any number below 7

    (c) increases

18. (a) Gas
    or
    Coal
    or
    Peat
    or
    Natural Gas

    (b) (i) Gases or below 20°C

    (ii) Brown

19. (a) (i) 5

    (ii) Slower or not as fast

    (b) Burns with a pop
    Burning/lit splint pops

20. (a) (i) C_4H_6
    H_6C_4

    (ii) aromatic

    (b) Any correct full or shortened structural formula with 5 carbons and 12 hydrogens

    (c) Bromine decolourises

21. (a) (i) Electrolyte

    (ii) zinc   copper

    wires

    (b) Chemical reaction stops
    or
    Runs out of chemicals
    or
    Reactants used up

CHEMISTRY GENERAL 2012

PART 1

1. (a) A   (b) D and F   (c) B and E
2. (a) B and F   (b) D
3. (a) C and F   (b) D
4. (a) F   (b) A   (c) B
5. C
6. (a) A and E   (b) E
7. (a) F   (b) C   (c) E
8. (a) C   (b) B
9. A and D

PART 2

10. (a) nucleus/nuclei

    (b) halogens

11. (a) Haber

    (b) Unreactive/Less reactive/Low reactivity
Not very/highly reactive
Does not react (easily)

    (c) Steel/brass/bronze/solder (or any other acceptable alloy)/pewter
Gold – must mention specific named alloy eg 9ct, 18ct but not 24 ct
Cast iron/sterling silver/dental amalgam

12. (a)

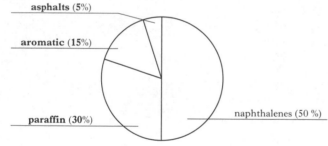

asphalts (5%)
aromatic (15%)
paraffin (30%)
naphthalenes (50 %)

    (b) Alkane(s)

    (c) C_nH_{2n}
$H_{2n}C_n$

13. (a) (i) A
       (ii) < 5 BUT > 0

    (b) Hydrogen/H^+

14. (a) styrene/monostyrene

    (b) Polymerisation/addition polymerisation/addition

    (c) Renewable/biodegradable/or any definition of biodegradable/will not run out/infinite source
Does not use up finite resources
Unlimited source
Sustainable
Can be regrown
Degrades
Not made from fossil fuels (oil etc)

15. (a) (i) chlorophyll

       (ii) Benedict's or Fehling's reagent turns brick red/orange/brown
(Must have reagent **and** colour correct)

    (b) Volume of water/mass of water
Mass of carbohydrate/amount/quantity/mass
Distance between spoon and test-tube
Particle size
Position of thermometer
Same test-tube/ size of test-tube

    (c) (not sweet) **and** (does not dissolve)

    (d) Oxygen/O/O_2

16. (a) Electron/e

    (b) Any metal below zinc in electrochemical series

    (c) (i) More power/current/voltage
Lasts longer/does not have to be replaced as often
<u>More</u> portable
Can power large devices/items
Can produce smaller batteries
Do not need to recharge battery as often
Do not use as many batteries/less waste as do not need to throw out as many batteries

       (ii) Li_2O

    (d) Vertical scale + label (½)
Correct bar labelling (½)
Bars drawn correctly (1)

17. (a) Nitrogen/N_2 and oxygen/O_2

    (b) Lightning/electrical storms

    (c) A given value less than 7

18. (a) Lattice/network

    (b) Ions not free to move/flow

    (c) colourless or no colour

    (d) Carbon/C
Carbon monoxide/CO
Coke/charcoal/graphite

19. (a) sodium carbonate

    (b) (i) 20°C ± 1
       (ii) As the temperature { increases / decreases }
the solubility { increases / decreases }
The solubility { increases / decreases } as the temperature { increases / decreases }

20. (a) (i) Cracking/catalytic cracking/thermal cracking
       (ii) Water/H_2O/hydrogen oxide/steam

    (b) It is unsaturated/not saturated
Contains carbon to carbon double bond
Contains   C = C
Contains double bond

    (c) Lead iodide